COSMIC
RELIGION

COSMIC RELIGION

An Autobiography of the Universe

Konstantin Kolenda

IMAGE BOOKS
DOUBLEDAY
NEW YORK LONDON TORONTO SYDNEY AUCKLAND

To Pauline

An Image Book
PUBLISHED BY DOUBLEDAY
a division of Bantam Doubleday Dell Publishing Group, Inc.
666 Fifth Avenue, New York, New York 10103

IMAGE and DOUBLEDAY are trademarks of Doubleday,
a division of Bantam Doubleday Dell Publishing Group, Inc.

This Image Books edition published September 1991
by special arrangement with Waveland Press, Inc.

Library of Congress Cataloging-in-Publication Data
Kolenda, Konstantin.
    Cosmic religion: an autobiography of the universe/
    Konstantin Kolenda.
        p.   cm.
    Reprint. Originally published: Waveland press, 1987.
    1. Man.   2. Cosmology.   I. Title.
BD450.K6394   1991   91-17433
113—dc20   CIP
ISBN 0-385-41962-7

8.00

Acknowledgments

At various stages of working on this book I have profited from comments and suggestions by people who read either parts or the whole of it. André Bacard, James Street Fulton, John Kekes, Pauline Kolenda, Margaret Miles, and David L. Norton have given me their helpful and sometimes critical thoughts, making the book better than it otherwise would have been, but I do not claim that they endorse every single idea in it. An early version was presented in a course in Continuing Studies at Rice University, and the response I received in that context encouraged me to think that the topics under discussion are of serious interest to people in many walks of life.

My special thanks go to Olivia Orfield, whose painstaking and meticulous copy-editing, combined with genuine support and encouragement, made the completion of the work so much easier.

Contents

Preface   xi

CHAPTER ONE
_____

MY PRE-PERSONAL EXISTENCE   **1**

My Star-Stuff Stage, 1

I Discover Life, 3

    The Novelty of Feeling, 3
    The Trick of Reproduction, 5
    The Profligacy of Nature, 5
    The Genius of the Genes, 7
    The Rudiments of Purpose, 8
    Pleasure, Pain, and Value, 8
    The Birth of Goodness, 11

I Become Human, 12

    Cooperative Arrangements, 12
    Tools and Language, 13
    Teaching and Learning, 14
    Rationality, 15
    Language as Instrument of Action, 16
    Culture, 17
    Refining Goodness, 19
    The Apex of Life, 21

CHAPTER TWO

PERSON: MY FAVORITE MODE OF BEING   23

From Species to Individuality, 23

    Self-Reference, 23
    Personhood, 25
    No Group Mind, 26
    Personal Judgment, 27
    Persons: Bearers of Culture, 28
    Tasks of Personhood, 29

From Nature to Morality, 30

    Caring, 30
    Morality, 32
    Prudence, 33
    Other Quasi-Moralities, 34
    Demandingness of Morality, 35
    Evil and Crime, 36

From Integrity to Autonomy, 38

    Moral Integrity as Achievement, 38
    Autonomy as Self-Development, 39
    The Scope of Autonomy, 40
    Obstacles to Autonomy: Elusiveness of the Self, 42
    Obstacles to Autonomy: Scepticism of the Psyche, 44
    Autonomy as Self-Construction, 46

From Meaning to Mysticism, 47

    The Birth of Meaning, 47
    Through the Eyes of Science, 48
    Qualitative Standards, 49
    The Fruits of Understanding, 52
    Piety, 53
    Mystical Feelings, 56

From Happiness to Blessedness, 58

    Personhood and Self-Conception, 58
    Happiness and Education, 60
    Happiness and Temporality, 61
    Blessedness: Affirmation of Value, 62
    The Pursuit of Blessedness, 63
    The Drama of Personal Destinies, 64

CHAPTER THREE

BARRIERS TO PERSONHOOD 67

General Limitations, 67

Absence of Enabling Conditions, 67
The Drag of Tribalism, 68

The Trap of Social Determinism, 70

"Society": A Weasel Word, 70
Socialization: Not Just Conditioning, 71
Community of Interpretation, 72

The Sex Differential, 73

The Myth of Aristophanes, 73
Sex and Culture, 74
Sex and Love, 76
Women as Moral Teachers, 77
Personal Equality of Genders, 79

The Ethnic Mask, 80

Ethnicity and Culture, 80
Ethnocentrism vs. Universalism, 81
The Common Humanity of Persons, 84

Ideological Warfare, 86

Two Social Ideals, 86
The Costs of Ideological Rigidity, 88
The Path Toward Accommodation, 90
The Personal Factor, 93

CHAPTER FOUR

MY GREATEST DANGER: 97
DEMISE OF PERSONHOOD

The Shocking Thought of Extinction, 97

The Message of Nuclear Silos, 98

My Future: In the Hands of Persons, 99

CHAPTER FIVE

MY RELIGIOUS QUEST   **105**

Personhood in Religion, 105

> Ultimate Concern, 105
> The Roots of Religion, 107
> Creator: The Great Agent, 108
> The Loving Sustainer, 110
> The Redeemer, 112
> Divine Judgment, 113
> Immortality: The Crown of Personhood, 116

Religion: My Cosmic Evaluative Quest, 119

> The Common Core of Religion, 119
> The Secular-Sacred Continuum, 122
> Practical Consequences of Cosmic Religiousness, 123

Postscript   **129**
Notes   **133**
Index   **135**

*To see the Earth as it truly is, small and
blue and beautiful in that eternal silence
where it floats, is to see ourselves as
riders on the Earth together, brothers on
that bright loveliness in the eternal
cold—brothers who know now that they
are truly brothers.*

— Archibald MacLeish

Preface

If the universe could talk, what would it say about itself? The question
sounds strange, and yet intriguing. It is strange because we know that the
universe cannot talk. The very idea of cosmos as an entity is philosophically
suspect. The cosmos *contains* things, but is not itself a thing. The tendency
to think about it as one huge all-encompassing entity is confusing because in
the very attempt to encompass it in a thought that thought places itself
outside the object it is about. To think about anything, even about a tiny
atom, is to be in some sense outside or beyond the item thought about. But
we can arm ourselves against the confusion by a simple reminder that the
universe we can think about is short of only one item: the thought which
thinks it. This does not seem like a serious deprivation; everything else is left
intact and includes whatever the universe in fact contains.

Let us, then, turn to the intriguing aspect of the question: "What would
the universe say about itself if it could think and talk?" Well, it would have
to tell us. But since it contains everything, including its history, it could tell
us many things in which it would be difficult to work up any interest at all.
It could tell us, for instance, how many grains of sand there are in the Gulf
of Mexico. Who would care to know that fact and to what purpose? So it
seems that what we want to know about the universe is not just any facts
about it but *significant* facts. The reason why it would be intriguing to hear
what the cosmos would say about itself is that, being all-encompassing, it
alone would be in a position to report what in its total expanse of space and
time is *worth* reporting. This is why we would want to eavesdrop on the

cosmos if it has a story to tell. If it does, what would it choose to tell us? What features of itself would it emphasize and what would it ignore? What is it proud of and what would it prefer to forget?

Because such questions intrigue us, it is tempting to resort to an autobiographical device and to let the universe speak for itself, in first person. This book unabashedly falls for this temptation and offers an imaginary autobiography of the universe. The temptation to assume the voice of the cosmos and to speak philosophically for it is not a sign of narcissism or megalomania. Rather, the point of embracing the autobiographical mode is to encourage the reader to pay attention to some objective features of the cosmos. Of course confirmation of whether these features are indeed objective and significant can come only after examining the contents of the book. My hope is that as the autobiography's philosophical "narrator" I managed to eavesdrop on things that the reader will find worth taking note of. If so, if the story unfolded here is plausible, the reader is likely to forget that the story has a narrator but will become absorbed in the story itself.

In resorting to the autobiographical device, I do not claim some special, idiosyncratic insight but on the contrary mean to call attention to features of the cosmos accessible to anyone. My point is that you, the reader, could have written this story as well. It is *your* autobiography too. Its main message is that from the very moment the universe discovered life it has embarked upon adventure the latest chapter of which was the emergence of persons. What I want to say, if only metaphorically and anachronistically, is that the cosmos from the very start *wanted* to become a person-bearing reality. And it finally made it! — in you, in me, in every human being. Although differently and uniquely expressed in each person, the universe lives out its quest for value in individual human manifestations.

It is important to drive this point home at a time marked by a growing sense of bewilderment, loss of nerve, and steady drift into scepticism or its desperate counterparts — narcissism and fanaticism. These unnerving tendencies of our time are unwarranted. When the phenomenon of life is placed in its total cosmic context, and when its path is traced from early origins to the present status in human form, something astounding and glorious is disclosed. By being attentive to the actual workings of life in us and around us, we can be led to appreciate and to affirm values that make our lives meaningful and vibrant.

When we examine our place in the universe, as this "autobiography" tries to do, we shall not think of ourselves as peripheral, accidental flukes in the scheme of things, but as entities which make this scheme worthwhile and meaningful. To have meaning and to be valuable, the universe needs beings that can make these features fully manifest. This book shows that human persons are such beings. But if so, each person can consider the account offered here as his or her story, which of course gains concrete content when it is filled our by the details of the individual's life. Indeed, the objective of writing this book is to restore to persons the crucial sense of their own dignity and self worth.

The quest for meaning and value is most insistently and tellingly revealed in religious interpretations of reality. The history of human cultures displays a rich variety of such interpretations. The tendency to see human reality in religious terms gives witness to the fact that certain values of life are perceived as having special significance. An examination of these values will disclose that they have their origin in special experiences of persons. Since these values cut across most religious traditions, it seems possible and desirable to think of religiousness as a universal phenomenon. Beneath the particularistic traditional forms there is a common core of religiousness which gives the quest for value a cosmic relevance. To make such a claim is not to indulge in grandiosity but merely to acknowledge the fact that human beings are the only inhabitants of the cosmos who can exhibit what Paul Tillich called "ultimate concern." To express such a concern in terms that do not limit themselves to any particular religious tradition is to acknowledge the possibility of a cosmic religion. The upshot of this "autobiography" is that such a possibility indeed exists and that it can be rationally defended. The bias toward a cosmic religion is not just a leap of faith but can be supported by a philosophical argument.

Konstantin Kolenda
Rice University

COSMIC
RELIGION

1

My Pre-Personal Existence

❧ My Star-Stuff Stage ❧

I am very old. Looking back at myself through the eyes of contemporary science I realize how dull most of my existence has been. There is really not much to say about it. The likely story is that I was born with a big bang. The form of matter in which I now exist was compressed to an almost infinite degree. But it seems that such a compression could not persist throughout eternity. In a dialectical fashion it contained within itself its opposite. At a moment which some may want to call the moment of my creation the infinitely dense mass of my primal substance exploded violently, and all heaven broke loose. The exploded matter receded from the center of the big bang, resulting in the gradual emergence of nebulae, galaxies, and stars, in varied and diverse configurations. The so-called laws of nature, with their arbitrary constants, imparted to matter scattered through space the bewildering configurations which astronomers now can see through their sophisticated telescopes.

Exemplifying the emerging laws, my matter in the star stage gradually defined itself as composed of elements. Each element comprises a different composition of particles, and each particle binds an incredible amount of energy. That energy displays itself in the cosmic spectacle of burning stars, streaking comets, and scattering radiation. The cumulative behavior of elements results in extremely high temperatures inside the stars, while the huge spaces separating stars and galaxies are incredibly cold. If I were asked to describe my state of being in the star stage in terms of feelings (which actually are attributable only to *living* beings), I would have to single out the contrast between feeling very hot (at the center of stars) and feeling very cold (in the interstellar spaces). But if I *may* be allowed to speak of feelings, in a metaphorical vein, of course, I should also mention the fact that the very being of matter out of which I am made up is constituted at its elementary level (the level of elements) by a sort of mad, intoxicating dance of infinitesimally small particles around their atomic nuclei. Were they life-like, they would suffer from chronic dizziness.

Relying on sophisticated atom-smashing laboratories, contemporary scientists are discovering that the composition of my elementary particles is extremely complicated, but they are pleased to discover that even these tiny particles are subject to strict laws, at least up to a point. This qualification is in order because physicists are compelled to acknowledge the presence of what they call quantum leaps in the behavior of matter at its elementary levels. Moreover, they run into conceptual difficulties when, for instance, they seem unable to ascertain at the same time both the position and the mass of my elementary particles. Be that as it may, and whatever scientific account will prove to be most plausible, I must confess that my existence in the state of particles speeding through space in various formations, thus expanding the universe in all directions, is a rather boring state of affairs, especially if you consider the time spent on these repetitive cosmic comings and goings. Although galaxies and stars travel through space at very great speeds, they are separated from one another by distances measured in millions and billions of light years. For an earthbound creature, the dimensions of concentrated matter, harboring fantastic energy and speeding through almost infinite expanses of space, may evoke a sense of immensity and power, but for me the spectacle loses its luster under the crushing weight of its sheer monotony. Somehow all that hurly-burly, this mindless (literally!) rushing through space—exploding stars, congregating galaxies, streaking comets, and the relentless spreading of ubiquitous radiation—after a while becomes old and dull. Apart from providing a potential spectacle for human beings, I would not be worth bragging about if this were all that could be said of me. Luckily, at some point of my long and boring history something different began to stir in me.

❧ I Discover Life ❧

The Novelty of Feeling

The novelty of which I am about to speak came into being at an unexpected place. Some of my innumerable galaxies contain solar systems. Such a system consists of a sun surrounded by planets orbiting around it at various distances. How such planetary systems get originated is a subject of lively scientific theorizing, one hypothesis being that the gravitational pull of a body passing in a sun's vicinity may separate a part of that sun from its center and hurl the separated matter into orbiting paths around the sun. Depending in part on its distance from the sun, each planet generates its own special physical conditions.

The conditions of one such planet, called Earth, were conducive to the emergence of something genuinely new. The interaction of elements out of which I am made up produces stable combinations called molecules. Some molecules, because of internal electric and magnetic forces, break up when brought into contact with other molecules. As a result new substances come into being, each manifesting different physical and chemical properties.

It so happens that the conditions of earth allowed one particular element, carbon, to form most intriguing alliances. When combined with other elements, such as oxygen and hydrogen, it forms molecules with unusual properties indeed. Their behavior is so different from everything else going on in me that it needs a special name: *life.*

The genesis of life on earth is not a settled matter and is subject to a vigorous debate. (My authority here is no greater than that of biologists who study these matters.) What is reasonably sure is that a combination of some molecules produced a highly reactive substance called protoplasm. It may be that protoplasm was formed in a "cosmic soup" of jostling molecules, tossed about by electric storms during which various chemical substances, especially amino acids, came to interact with one another. Countless combinations succeeded one another until some of them took on features propitious for life manifested in protoplasm. Under these special conditions some intermediary entities, such as viruses, straddling the boundary between living and nonliving matter, could also come into being.

Protoplasm formed itself into entities which the biologists have dubbed "cells." Cells display a very special capacity: they can absorb from their surroundings substances which they incorporate into their own make-up. The wonderful result is that cells, by virtue of their assimilationist capacity, can maintain their integrity as persistent entities. This capacity to sustain themselves as internally-organized units justifies calling them by a special name: organisms. The materials which an organism uses for self-maintenance are appropriately called nutrition or food. Since the process of absorbing nutrients from the surroundings is easier in some physical media than in others, biologists speculate that my first organic or living cells were formed in water, a substance in which particles usable as food are suspended.

I must confess that in a living cell I could experience something unprecedented in all my previous existence: I discovered what it is to be reactive, actively responsive to immediate environment. A cell is kept alive by "welcoming" into itself substances which it can absorb. It manifests something that a philosopher named Spinoza was to call *conatus,* a desire for self-preservation. *Conatus* is responsible for the cell's assimilation of nutritious particles and for its turning away from and avoiding contact with substances that are useless or harmful. It seems a simple matter, and yet how revolutionary! It introduces a radical novelty: *feeling.* In order to discriminate between food and non-food, the cell must be able to exist in at least two different inner states, a welcoming one and a rejecting one, so to speak. Of course these states must be understood to exist in a rudimentary, elementary sense. The descriptions I am using here have been coined by much more complex organisms, human beings (to whom we'll come in due time), and therefore have much richer connotations than the simple and primitive one that arises at this early stage of my living career.

To characterize the revolutionary change from non-living to living states, I am almost tempted to invoke a notion of special importance to a later development, religion: the notion of miracle. No other notion seems able to account for the transition from a situation in which entities existing side by side are wholly indifferent toward one another to a situation in which the surrounding of an entity begins to *matter* to it. To a living organism its surrounding is far from being a matter of indifference; it is literally of vital importance! For it may signify either survival or demise. I must apologize for smuggling in at this point the notion of "significance"; of course it is used anachronistically. But my apologizing here is a bit half-hearted. With the advent of possible discrimination between welcome and unwelcome states there is some justification for using the notion of meaning—again in an elementary, tentative sense. I shall return to this point soon.

I cannot overestimate the import of discovering in myself this new possibility. It was quite exhilarating and disturbing. If by "disturbance" is meant "agitation," then this description is also literally true, because being alive *means* to be agitated, reactive. A living substance is of course material, but it is no longer content to remain locked into its inert, self-sufficient, essential being—as is an atom or a molecule or even a huge collection of them, that is to say, a star. As Jean-Paul Sartre was to characterize such a "dead" entity, it is an *in-itself,* in contrast to an entity whose very being depends on "taking account" of its surrounding, which exists *for* it. By assimilating selected parts of that surrounding, and by eliminating a used-up (digested) or unusable part of it, a living being is *animated* in the most basic sense of the word. Feeling is the internal stage of an organism as it brings about these transactions.

The Trick of Reproduction

Most early forms of my life were unicellular, i.e., consisting of one cell. To my amazement I came to discover that even a divided cell could maintain itself and remain an integral, viable entity. In effect, by cell division more surfaces were created to bring the organism into bartering transactions with its environment. *Conatus* really took hold! But that's not all. Protoplasm not only could reproduce itself by cell division; it could also differentiate its cells into specialized ones capable of performing particular functions. Some cells became *sense* organs, especially sensitive to such phenomena as light or sound waves. These phenomena produced *sensations,* which inform the organism about some features of its surrounding. As a result, the organism was no longer limited to undifferentiated feeling, uniformly spread along its simple body. It could experience varied agitations of different degrees of intensity in different parts of the now more complex body. The activity of sense organs transformed feeling into sentience.

All of this of course took time, but eventually living cells could "grow" into bigger, more complex organic structures. Naturally, the increasing internal differentiation was accompanied by a greater and more variegated reactiveness. A more complex organism can experience feelings that are "richer" than those of a simple one, and it is clear that as this capacity enlarged, the contrast between life and non-life became more prominent. That contrast became even sharper when, besides developing functionally useful differentiations, some organisms delegated the task of reproduction to special organs. At first the sexual function was undifferentiated, but some organisms transformed themselves into reproductively complementary entities, thus giving rise to genders. In such organisms the possibility of reproduction depends on collaboration between the male and the female versions of the species. This process was a tremendous breakthrough, allowing the sex differential to take a great variety of forms. I must say that experimenting with them introduced much additional zest into my living. *Conatus* with a vengeance!

The Profligacy of Nature

The self-propelling desire captured in the Latin noun *conatus,* calls up another more familiar word, *nature.* The root of this word is the verb *nascere*, "to be born." Having once been born, life uses birth as a device to have more of itself. I must admit that my appetite for life is insatiable. Starting with plants firmly fixed on one spot, life forms gradually developed ways of moving their locations, thus literally giving rise to the locomotion of animals. This alteration was brought about by the development of a great variety of special locomotive organs—fins, legs, wings. To be sure, the changes in living organisms, in their structures and functions, were in part prompted by changes in atmospheric, meteorological, and geological conditions of the planet. I am not ashamed of the tenacity with which I threw myself into the *conatus* of life, making its survival my paramount objective. Anyone familiar with the history of life on earth knows how determined and inventive I have been in promoting my newly discovered (as compared to the eons of time when my existence was literally lifeless) biological realm. My creative exuberance did not escape notice and has been frequently remarked upon. Annie Dillard is quite right when she says in her delightful book, *Pilgrim at Tinker Creek:*

> Nature is, above all, profligate. Don't believe them when they tell you how economical and thrifty nature is, whose leaves return to the soil. Wouldn't it be cheaper to leave them on the tree in the first place? This deciduous business alone is a radical scheme, the brainchild of a deranged manic-depressive with limitless capital. Extravagance! Nature will try anything once. This is what the sign on the insect says. No form is too gruesome, no behavior too grotesque. If you're dealing with organic compounds, then let them combine. If it works, if it quickens, set it claquing in the grass; there's always room for one more; and you ain't so handsome yourself. This is a spendthrift economy; though nothing is lost, all is spent.[1]

Dillard's comment on insects applies to all of my creatures, and insects are but a small segment of my biological adventure. Consider the diversity of life forms that flourished on earth in the past. Allow your thoughts to range over the entire spectrum of animal orders that have inhabited this earth across the evolutionary eons. Encourage your imagination to tell you what it might be like to be an insect, a reptile, a bird. Imagine the perceptions and sensations that are felt by creatures endowed with a vast variety of sense organs, moving within environments quite different from the present one, seeking food, shelter, and avoiding harm and danger. What must it have been like to be a dinosaur feeding on abundant plant life, attacking a mating rival, or seeking relief from burning sun in a cool lake? A generous imagination is required for such speculation, and I myself have but a dim memory of the kinds of sentience which agitated me across eons of time. What is known to contemporary scientists is only a fraction of that plenitude of life forms which my cosmic urge tried to explore. I don't deny that countless species flourished only briefly and were left behind in dead corridors of unrecorded time. I am not unfamiliar with extinction.

The Genius of the Genes

Although many species of flora and fauna have come and gone—their *conatus*, the desire to preserve themselves, frustrated by adverse conditions—life had no difficulty in generating other genera and species. How this great chain of living being forged its continuous links is of great interest to biologists, and they have come to learn much about the cunning mechanisms my living substance utilized in order to keep itself going. Even though the biologists' findings are still far from complete, they have unearthed a great deal about my survival-enabling traits. Darwin discovered that in the struggle for survival some organisms were aided by fortuitous mutations that bestowed on them special advantages. Some mutations were caused by cosmic rays constantly bombarding the earth. Fortuitously occurring, mutations are preserved in subsequent generations through the action of special cells called genes, which determine the inheritance of reproduced offspring. In addition, deviations from the organism's original structure, either in a negative or a positive direction, can be attributed to the malfunction of the information system programmed into the cells. As an organism develops, its cells may be said to "read" a certain score, stored up in the organism's genetic make-up. When the programming mechanism is disturbed structures and corresponding functions deviate from formerly established patterns. They may go awry in various "monstrous" ways, or, alternatively, they can make the organism more viable.

Genetic cells appear to "guide" the organism's development, as if they "knew" what is good for it. But I should warn against reading too much into my activity at the pre-human level of life. There I do not display what has come to be characterized as "purpose." The structures and functions exhibited by organisms at pre-human levels of life can be said to be purposive only in an extenuated, minimal sense. Biologists are quite right in saying that as long as the behavior of DNA cells is governed by definite chemical reactions the language of purpose is dispensable. When I ask myself why genetic cells behave the way they do, or more generally, why in the long process of evolution some forms of life flourished and others did not, I don't have an answer. I can only say, retrospectively, that some forms happened to survive. Like the biologists, I can "explain" the survival by merely pointing to the features that proved to be important for survival. It was just a matter of *conatus* plus luck. I know that the answer is disappointing, but it cannot be otherwise because at that stage the notion that I was indulging in what came to be known as evolution could not even occur to me. I was simply enjoying the feelings that emerged in the organisms as they defined themselves out of fortuitously-arranged circumstances of the physical conditions of the earth. As Annie Dillard remarks, if something worked, it was good enough for me.

The Rudiments of Purpose

It is true that the success of life forms in maintaining themselves produced in me welcome reactions. So, it is not incorrect to say that at least to that extent a *purpose* was realized. The reason the use of purposive language must be carefully qualified when speaking of pre-human levels is that, as we shall see, the full-flown notion of purpose has found its true home much later, namely, in the human context. Still, the very fact of *conatus*—the desire to preserve one's life—can be said to contain the germ, the minimal sense, of purposiveness. The welcoming, life-affirming feelings experienced by an organism *lay the ground* for the phenomenon of purpose. An amoeba veering away from a sharp protrusion in its environment can be said to move purposely. Its movement has a point, a rudimentary meaning or significance; by avoiding the protrusion it is able to preserve itself. Much differentiation and complexity in the make-up and behavior of living organisms took place before the notion of purpose became fully applicable, but it is undoubtedly true that in discovering life I was rapidly moving in the direction of full-fledged purpose.

True enough, the transition from purpose taken in an elementary, attenuated sense to the fully-developed one took a very long time. But the time factor is not really important. After all, I have an unlimited supply of time. Although my cosmic energy at the level of elementary life was literally thoughtless, still, in virtue of its capacity for feeling, it already had a way of directing itself toward success or viability, without even aiming at it. But since *conatus* is continuous, present both in simple and in complex organisms, I am not unjustified in reading the idea of purpose back into my life's origins. The development of life constitutes an unbroken chain, and all of its forms are bent on survival. In *homo sapiens*, the most recent link of that chain, this objective takes on the character of explicit purpose, and the continuity of that objective entitles me to some retroactive descriptions of what was really going on.

Pleasure, Pain, and Value

There is still another reason for attributing a basic purposiveness to the entire chain of living beings. The distinguishing feature of life—feeling or sentience—has two basic forms: pleasure and pain. These two opposing poles of feeling, experienceable in various ways by all living beings, testify to the continuity of purposiveness in the minimal sense as attraction to pleasure and avoidance of pain. Jeremy Bentham was exaggerating when he said that "Nature has placed mankind under the governance of two sovereign masters, *pain* and *pleasure*," but his observation is essentially right when applied to pre-human forms of life. Organisms seek to maintain and to prolong pleasant experiences and try to escape or to terminate painful ones. From the noises they emit, one can tell when dogs or cats are experiencing pain. Simple animals such as worms lack the requisite nervous structure to experience something analogous to what is experienced by higher animals, but their behavior justifies the supposition that they do not welcome a destruction or injury of the tissues that make them up. Just look at the wriggling worm when you prod it. Although the sensations of pleasure and pain depend on the physiological make-up of an organism, all living beings possess "inner life." To say this is to indicate that they react to the effects of their environment, that they experience satisfactions and discomforts.

Of course, the very idea of *experience* has its meaning only within the phenomenon of life. It must be put alongside other notions I needed to invoke in order to characterize the mode of being alive. Among them first of all is the notion of feeling or sentience. Following close behind are the notions of significance or meaning, understood in the elementary sense of a capacity to distinguish between friendly or hostile, welcome or unwelcome occurrences. As noted earlier, the notion of purpose, also in its rudimentary sense, was being born here—as the tendency to react in ways that realize survival and avoid destruction. Even the notion of intelligence may be invoked at this stage when it is understood as the ability to make discriminations which bring about self-preservation. To embrace *conatus* is to find something to be attractive. Thus *life is my primordial act of affirmation.*

I must emphasize that there was no room for the notion of affirmation in my pre-life stage. In my star-stuff stage, at the levels of my physical and chemical existence, there are no phenomena to which this notion could apply. At these levels the purely quantitative categories of energy, mass, speed, and size are sufficient to describe what is going on. But in discovering life I have found a completely new, qualitative dimension in myself, namely, the dimension of *value*.

I must make a further confession. My newly encountered living mode of being became for me a source of narcissistic fascination and self-absorption. Once *conatus* got hold of me, I found myself eager to give it full reign. The impulse to perpetuate itself is so strong in living beings that, to explain it, some philosophers postulated a fundamental metaphysical principle: life force or *élan vital*. One concomitant of this irresistible surge of life in me was its blind indifference to the way the survival is assured. It turned out that the quickest and most effective way of obtaining food was to devour other organic substances. As a result, in the state of nature life feeds on itself. In what is called the ecological system one form of life consumes other forms. This competition for food speeded up and diversified the process of evolutionary change. It set in motion the struggle for existence in which the fittest survive and the weak perish. I must confess that the spectacle of life devouring life has not become problematic for me until I have reached the human stage, when such pejorative slogans as "nature raw in tooth and claw" were formulated. Just how and why such slogans emerged is a long and fascinating story about me, still to be unfolded in this autobiography. At this point I must limit myself to admitting that in my irresistible zest for living I was oblivious to the self-devouring character of my mode of being. It testifies to the great intensity and enthusiasm with which I threw myself into the exploration of seemingly endless variations on the theme of life. The love of life in me was so great that I pursued it headlong without regard that to an onlooker it might appear to be a war of each against all.

I must qualify this observation by saying that the very fact of struggle and competition resulted in many species developing defensive modes of behavior — protective coloring and effective mechanisms of attack, resistance, and escape. Mutations, fortuitous adaptations, and favorable physical conditions allowed many species to prevail while others disappeared from the scene. The fact that life forms proliferated on earth testifies eloquently to the zest which I have manifested in this mode of being. Only much later could the question arise as to whether the whole biological realm could display itself in some other, possibly preferable way.

But even before reaching the human stage my many life phenomena flourished exuberantly, and still continue to flourish. I want to emphasize that my *conatus* is not limited to the desire for self-preservation; it *delights* in its mode of being and seeks to enlarge its domain. It is distinctly experimental in character, always bent on displaying itself in new, refined variations. The character of refinements of which my life is capable was aptly described by another sensitive observer of life phenomena, the biosophist Stephen Lackner.

> What is meant here by refinement? It is the difference between the sala-
> mander groping through mud, thick, vulnerable and clumsy, and the
> slender, firm, agile, and graceful lizard; between the scaly, unadorned,
> flowerless liverwort and the showy, highly differentiated orchid; between
> the swarthy, phlegmatic tapir and the nervous, spirited horse; between
> the physically overpowering, shaggy gorilla and the highly trained, ele-
> gant airline pilot.[2]

This description moves ahead toward the human stage, thus anticipating my later discoveries to be discussed shortly. But I must also agree with Lackner when he suggests that the tendency toward beauty was present in me from the very start of life creation, which was restless and adventuresome. In seeking more useful forms, my life tends to prefer those that humans consider beautiful, as the glorious colors of butterfly wings and the multi-tinted plumage of birds eloquently testify. The esthetic element is also present in elaborate mating rituals, giving some justification to the claim that often the survival of the fittest was also the survival of the fanciest. Many a bird indulges in song for no utilitarian purpose, but only to express the pure joy of aliveness. Living existence is much more interesting than the inertia of physical matter. That's why the discovery of life, in the narrow zone of temperatures between the freezing and boiling points of water which happily prevails in this small corner of my vast domain, was such an important event in my cosmic career.

The Birth of Goodness

When I compare my star-stuff stage to the living one, displaying itself in one tiny corner of my physical expanse, I am struck by the tremendous contrast between them. It is dramatic. If one recalls that the Greek word "drama" means "action," one is bound to realize that this name applies in both an evaluative and a literal sense. In its living form my basic matter is no longer confined to the repetitive dance of electrons around protons, accompanied by the confused swarming of even tinier particles the physicists are now discovering in their fancy atom-smashers. I can display myself in substances that are not indifferent to what is happening to them, that find their surroundings *interesting*—friendly or hostile, supportive or destructive. I cannot escape invoking here, for the first time, a notion that will take on an even greater prominence as I tell my story, the notion of *goodness*. I declare roundly that it is good to exist in the form of entities that are animated by a special inner mode called "feeling." To be animated is to be no longer inert, cold, literally dead, but instead to *care* about what is going on—both inside and outside oneself. When Martin Heidegger reminded the modern philosophers that the primary mode of being human is not cognitive awareness but care *(Sorge)* he was calling attention to something in human life that has its roots in the ancestral beginning of all of life. I should also add in this connection that in discovering life in its primary mode of care I have also discovered something about *myself*; life phenomena are not an intrusion from some mysterious supercosmic realm (after all, I *am* all that is). I am proud to display my cosmic energy in this literally vital form.

⚡ I Become Human ⚡

Cooperative Arrangements

The exuberant conquest of the earth by life in its variegated forms is a clear testimony that I do favor this form of existence; I must find it congenial. At least, it is a fact about me that the viability of life, under very special conditions, of course, is not in conflict with my essential reality. Life has turned out to be one of my real possibilities. The manner in which I am actualizing it speaks for itself—in the irrepressible, extravagant, and proliferating *conatus.*

My infatuation with life is borne out by another phenomenon, one that exists side by side with the competition among living beings for survival, the Darwinian struggle for existence. Prince Peter Kropotkin was one of many biologists who were struck by the *cooperative* arrangements of life. Such arrangements, both within and between species, are clear indications that the thrust of life crosses the boundaries between individuals. This thrust is acknowledged in the notion of social instinct. Members of a species develop mechanisms for cooperation and specialize in the division of labor, especially evident in the mode of life of such social insects as ants and bees. This tendency of nature prompted Goethe to observe that it favors the species over the individual.

It is again tempting to invoke here the idea of purpose and design. This concept would provide an explanation of the cooperative features of life. But here, too, I would like to disclaim any direct awareness of purpose on my part. Social cooperation and various forms of symbiosis can be explained on the evolutionary principle: species that developed cooperative schemes fared better than those that did not. The same can be said of other useful instincts, such as caring for the young or migrating to warmer climates for the winter. Again, I would remind the reader that in exploring my various possibilities time was of no consideration; I have plenty of it. At any stage of my temporal journey I do not display any more capacities than are manifested in the forms of life actually tried out.

Is life at pre-human levels *intelligent?* Yes, in the specific ways in which it actually *is* intelligent—clinging to those accidental mutations and symbiotic patterns of behavior that proved successful in the struggle against competitors and the changing environment. There is nothing wrong in calling instinctive adjustments and adaptations a *form* of intelligence. However, it is important to note that this kind of intelligence is attributable neither to individual members of a species (they don't think about mechanisms that literally move them) nor to me as the alleged sponsor of some metaphysical principle such as life force. In making life viable—in whatever form it found expression—I was working in the dark, so to speak, not equipped with some antecedent awareness or knowledge of what would work and what would not. I am *not* omni*scient*; the very notion of knowledge (*sciencia*) did not occur to me until I could manifest myself in human beings. But I cannot deny that this whole exercise in the vital arena absorbed my most creative and inventive energies available to me under the special conditions of the planet earth.

As I have noted before, the more complex living organisms became, the richer was their (and my) inner life. The kinds of feeling and sentience I could experience are a function of structures which the given organism develops. Sense organs, defense mechanisms, and reproductive apparatus kept enriching the nervous system of the organism, which is the internal framework of communication inside the body. In a being that in due time came to call itself *anthropos* in Greek and *homo sapiens* in Latin, the need to coordinate the multiple organic functions gave rise to an elaborate, intricately structured brain. The coordinating function of the brain had to include the management of two especially important new traits that emerged in the evolutionary process: erect posture and the hand with its opposable thumb. These traits made possible new ways of coping with the environment.

Tools and Language

Among the most prominent new possibilities was the use of *tools*. Unlike other animals, human beings no longer had to limit themselves to the natural, "built-in" features of their bodies—legs, hands, teeth—to look for food or to defend themselves. To perform many tasks they could now utilize also *artificial* means: sticks and stones at first and eventually almost anything in the external environment. Contemporary human beings can literally move mountains—by means of dynamite and bulldozers.

But even the tool-using capacity of human beings is overshadowed by another new mode of behavior: *language*. In one sense language is also a tool, an instrument put to special uses. A general term for one use is communication, but since there are so many different types of it, finer distinctions are called for. Some forms of communication are employed by non-humans as well. Many animals use sounds to express their feelings, attract mates, warn of danger, frighten an enemy. Sounds and bodily gestures perform many useful functions in pre-human organisms. A completely new function, however, comes into being when a sound is used to *refer*.

I cannot stress enough the importance of this new phenomenon. In time it revolutionized the entire biosphere and opened for me undreamed of horizons. Philosophers are still arguing vehemently over the most plausible account of what it means for language to refer. It is safe to assume, however, that at first the sounds uttered by ancestors of humans in the presence of some things, events, or animals were merely expressions of emotion: fear or delight. The ability to use a sound to refer to a thing did not spring into being in a pure form of direct reference. Sounds became signs or words probably in the context of vital urgency. A sound uttered in the presence of some danger may have been at first *forced* from an animal as an expression of fear or warning. But when besides this expressive function the sound came to *mean* what it refers to, a most significant shift took place. In the referential use the sound didn't just signify "I am afraid" or "Danger" but "*There* is a dangerous creature" or "There *it* is again."

Imagine the benefits for a species that could make use of this referential function of words in communication. How different is the behavior of a being who is afraid of bears and who can now understand the meaning of sounds corresponding to such situations as "bear in tree," "bear in the lake," "bear here," "bear there," "bear in daytime," and "bear at night." In general, a group of individuals attending to things of mutual interest—searching for food, fighting enemies, distinguishing among sources of concern, danger, or delight—can use language to initiate cooperative behavior that moves beyond the confines of merely instinctive reactions. Once established, the referential use of language can be transmitted (in contrast to instinct which is inborn) to offspring through *teaching.* Teaching and learning make it possible for individuals to move beyond the conservative inertia of instinctively built-in patterns of the species. The capacity to refer has been facilitated by the invention of tools. A tool-using being is bound to be struck by the *separateness* of the tool from one's own body; hence a divided attention, differentiating between the two items, is made possible. Furthermore, the attention can be focused on the tool itself, apart from its function, thus bringing into prominence the mental phenomenon of referring.

Teaching and Learning

The establishment of meanings depends on repetition. When language learners come to react to sounds in the same way under similar circumstances, sounds become *words*. In that sense, words are arbitrary, artificial creations of language-using animals. They are arbitrary in the sense that some *other* sounds could be given the particular linguistic function. The French do not make a mistake when by "pain" they mean "bread." What is important about language is that it makes it possible to realize values that otherwise would remain unsatisfied. If, as a human being, I like honey, for instance, my ability to satisfy my appetite for it is enhanced when I can be told where I can find it. A communication to that effect is an *additional* value. "Knowledge is power," said Francis Bacon. Hence, the ability to communicate knowledge through language is a valuable commodity. Use becomes useful.

An enormous consequence followed from the birth of the referential use of language. Human beings could now tell each other things, and tell them *correctly* or *incorrectly*. The one who has learned the meaning of the word "bear," for instance, has also learned what situation *justified* him in calling a creature by that name. With that, a notion of rightness or wrongness, correctness or incorrectness, truth or falsity was born. Human beings could now learn and teach words. In teaching and learning, proper uses of words are distinguished from improper ones. Thus *standards* of right or correct saying came into being.

The notion of correctness is clearly evident in the learning of language. A child gets certain verbal expressions either right or wrong. By various means, the teacher tries to make sure that at a certain stage the child distinguishes between rightness and wrongness, correctness and incorrectness. The process of getting the linguistic rules right is a complex one. It involves pointing, repetition, variation of context, and may be backed by reinforcements, punishments, and rewards. A human being who uses a word on his own manifests a mastery of a technique; he is able to "go on" in the same way under circumstances like those in which the teaching occurred.

Rationality

In the capacity to learn, the advance over instinct is enormous. Communication by means of language, based on the ability to apply common standards of correctness, can step in when the instinctual reaction breaks down and thus fails to secure a vital need. In its place comes a verbal message. With the emergence of language life makes a qualitative leap. I become capable of a new form of intelligence. That new form can be called *rationality*. Rationality goes beyond prelinguistic intelligence, since it involves the ability to recognize and to follow standards, artificially stabilized marks, at first sounds and later written signs, acquired through learning. This feature is absent from instinctive behavior, which is inborn and is governed by physiological structures. A very important shift takes place at this level of my experimenting with life. Even if one calls instinct a kind of animal intelligence, that intelligence differs from rationality because it does not consist in the *performance* of intelligent *acts*. (We shall return to this important point.) Instinctive "actions" are in truth no more than reactions; they do not involve a consideration and application of acquired standards. Instinct can be intelligent or unintelligent, but not rational or irrational. In contrast, actions can be appraised in both dimensions.

Here we have the great contrast between strictly animal and human ways of realizing values. Values brought into me by human beings are actualized with the help of a new tool—language—by deliberate actions of standard-using animals. In the human form of life, sentience and instinct have been supplemented and enhanced by rationality, which is language-based, standard-enacting intelligence. The transition from prelinguistic intelligence to language-based rationality was a qualitative leap, introducing a wholly new dimension into my cosmic existence. In that dimension life thrives not only in sentient and intelligent but also in rational creatures.

Language is a cooperative venture; it requires a community of speakers and hearers. To understand what it is to refer to an object is to understand that exactly the same reference can be performed by others, and this common understanding is the birth of a special kind of community. In language, items of experience begin to be commonly understood or interpreted. Thus for a human being a *common world of objects* emerges. When grasped by and articulated in language, the worldless sensations of sentient beings become *perceptions,* in which sensations are clothed in a form-giving, standard-bearing concept. In time, discriminations within the field of perception introduce into human consciousness the awareness of multiple relationships, of contrast and coherence, clash and harmony. Correspondingly, language becomes more complex and in addition to establishing the meaning of individual words (semantics) it acquires syntactic structures.

Language as Instrument of Action

So powerful is the referring function of language—identifying objects for which individual words stand and describing situations which a sentence depicts—that philosophical accounts of language, especially in the Western world, tended to regard this function as primary and tried to reduce to it all other linguistic functions. Only recently it has been shown, most powerfully by Ludwig Wittgenstein, that this view is reductionist. It distorts other important uses of language. Some verbal expressions, those of pain, for instance, do not describe anything but only replace non-verbal expressions of pain. Language also enables us to *bring about* certain states of affairs. A bride who pronounces the words "I do" at a marriage ceremony commits herself thereby to a life-long companionship. A judge saying in a court proceeding the words "I sentence you to a year in jail" deprives the defendant of his freedom. Language can be used to make promises, to lay down verdicts, commandments and prohibitions, to praise, to blame, to criticize, to curse.

As Wittgenstein observed, language is never *mere* language; it derives its importance from "the stream of life" in which it is embedded. It functions against the background of concerns and practices to which it gives voice and articulation. When language is characterized as a form of life, neither "form" nor "life" are subordinate to each other. The distinction between intelligence and rationality presupposes values which I sought to realize and maintain even before the phenomena of language came upon the scene in the human species. Rationality expressible in linguistic behavior is superimposed on the preexistent capacity of human beings to be intelligent in other ways. The prelinguistic intelligence exists prior to and side by side with the ability to master language in the process of socialization. That socialization no doubt includes an introduction to nonlinguistic skills and techniques and of nonverbal signs and gestures as well. Verbal communication is but one, although an extremely important, way of showing one's humanity. It supplements "body language" and countless other ways of conveying meanings through manifestations of pain and pleasure, through moods and emotions with their corresponding expressions.

Undeniably the explicitly verbal use of language has brought into the life of beings who discovered it — *homini sapientes* — a correspondingly rich flood of feelings and experiences. The attention to and discrimination among colors alone made them sensitive to the whole world of optical splendor. Likewise, the content and messages of other senses, when taken up into the resources of language, show new, previously unperceived relationships. As a result, both the external scene and the perceivers' inner lives become correspondingly enriched.

Culture

The qualitative leap toward language also made it possible for groups of humanity to transmit, perpetuate, and develop what has come to be called *culture.* Culture is a pattern of particular forms of behavior — practical, ritualistic, and expressive. It enables its participants to live in an *interpreted* world, a world full of meanings. Immense advantages resulted from the possibility of transmitting culture. By means of language, knowledge can be preserved, skills can be improved, refined, and expanded. The drawback of an instinct is that it remains rigid, unchanging. In contrast, *acquired* knowledge and skill can be modified in the light of changing circumstances.

Intelligence, transformed into rationality, led to an explosion of knowledge. Under certain auspicious circumstances civilizations emerged;

that is, the life of some large groups of people became sufficiently structured to enrich the social life with many specialized rules, observances, and rituals, and to organize it around a variety of tasks and activities. Special offices, obligations, and tasks were assigned to particular individuals, requiring initiation, learning, teaching, and apprenticeship. Specialization followed, dividing human activities into practical, political, and religious arts, in time contributing to the emergence of what nowadays is called fine arts, literature, and music.

Surveying the expanse of civilized human history, I can list among the distinguishing features of humanity diverse phenomena of practical and political arts: engineering and technology, religion and literature, philosophy and science, architecture and various fine arts, morality, and statesmanship. Culture is a system of standard ways of speaking and acting shared by members of a given community. Human history is the story of cultures, of living in the light of norms embedded in folkways and institutions. Cultural norms make possible the distinction between excellences and vices, values and disvalues recognized by particular human groups. Historians record the flourishing and the decay of tribes, nations, races, and civilizations.

It is a matter of no small importance that in the course of its history the human race has developed thousands of languages. A contemporary language atlas would show between four and five thousand that are distinct. Regretfully, like extinct species, thousands of these are dead, no longer in use. It would be a staggering, nay, impossible task to imagine the variety of uses to which language has been and can be put, and the variations of structure it exhibits. A mastery even of *one* language in its sophisticated reaches—in science, literature, and poetry—is a task in which perfection and revealing success are rare and often awe-inspiring. The moral of the story of Babel need not be necessarily negative. Although it may point to the drawbacks of mutual noncomprehension, it can also be seen as a tribute to the prodigious linguistic capacities of the human race.

Short and insignificant as the span of human history on earth may appear when compared to the age of this planet and of this solar system, it nevertheless exhibits an enormous wealth of standards and norms of excellence developed by human beings. For thousands of years this earth has been animated by the life of peoples who enacted the dramas of rite and ritual, of familial, tribal, and national hopes and aspirations, of artistic and scientific creativity.

Compared to the complexity, richness, and endless inventiveness of this distinctly human way of being, my star-stuff mode of existence appears dull indeed. The automatic, mechanical, and repetitive behavior of subatomic particles, and even that of biological cells, dazzlingly displaying the feats of

DNA chains, seems rather pedestrian, in spite of the intricacy of their structural relations and of the immense physical energy harbored at their centers.

In saying this I do not mean to slight my previous forms of life. I still marvel at the incredible interplay of structure and function in such special organs as eyes, for instance. I understand the reluctance of scientists to acquiesce in merely mechanical explanations of such biological feats; they seem to cry out for an acknowledgement of intelligent design or purpose. Impressive as artificially constructed modern computers are, they do not measure up to the naturally developed human brain. I am proud of my life at its instinctive level.

But that pride grows when I look at the accomplishments of human culture. It would be insensitive of me not to be impressed by the proliferation of artifacts and works of art, of scientific hypotheses and mathematical theorems, of social, moral, and political patterns and principles enacted by human beings in the course of their history. The attainments of the human spirit present a spectacle of its own kind and manifest values of an order altogether different from that which I exhibited in purely biological phenomena.

Refining Goodness

Rationality, a step beyond animal intelligence, is, nevertheless, a continuation of my experiments with value, with the quest of goodness. That quest seeks ever more refined, more enticing forms. At first guided by instinctive intelligence, the pursuit of goodness was intensified with the advent of rationality. The direct connection between rationality and goodness was evident to Socrates. He criticized Anaxagoras for claiming that things in the universe are governed by Mind but then failing to show that action of Mind was "for the good" of anything. If that which Anaxagoras meant by Mind operates mechanically, without invoking any standards of what is good, then Mind is literally mindless—a contradiction in terms. For an activity to be rational, it must aim at some value; that is, it must be guided by some standard in the light of which the activity appears worth pursuing.

I might mention in passing that the Socratic view can be invoked to settle the contemporary dispute as to whether computers are intelligent. If the distinction between intelligence and rationality is duly noted, there is no need to be squeamish about imputing intelligence to computers, or to speak of them as manifesting "artificial intelligence." Computers "perform" amazingly complex tasks, and are said to be doing it "on their own" if provided with sufficient "memory" and "skills." But they still fall short of rationality, when the latter is defined as the capacity to aim at some state of affairs regarded by the *computer* as worth bringing about. The most sophisticated computers produced so far have no aims of their own; they only help to achieve the aims of programmers, human beings. If the day should arrive when the computers develop their own independent aims, they would have to take into account the standards of values already developed in my human form of life, or at any rate the standards of computers would be subject to evaluation by *human* standards.

The distinction between intelligence and rationality also calls attention to what is to be included in a rational *explanation*. Such an explanation connects an act with bringing about what is valuable. If it can be said correctly that the reason for an event or action was the production of a state of affairs valued in accordance with some standard, we have produced a rational explanation. When such an explanation is compared with a *scientific* explanation, in which an event is said to occur in accordance with some law of nature, it is clear that the scientific explanation lacks a reference to what Socrates believed to be a necessary component of a rational explanation: satisfaction of some good. Socrates found Anaxagoras's explanations unsatisfactory precisely for this reason: Anaxagoras's Mind merely manifested the operation of some value-neutral laws. This is what natural science does; it explains a phenomenon by invoking the operation of a law. When the question is asked how the law itself is explained, a scientific account tries to subsume it, if possible, under a more inclusive law, and so on, without ever raising the question of what use are these laws. But unless this question is raised and answered, explanation is doomed to infinite regress. In contrast, rational explanation justifiably can stop at a point when an activity is shown to result in something valuable by reference to some standard. The word "standard" itself has a value connotation. A standard provides a yardstick, a measure of determining whether something meets certain expectations of the desirable.

In making these parenthetic remarks, I do not mean to disparage science. As actually employed by human beings, the procedures of science exhibit both intelligence and rationality. Although scientific explanations are value-neutral, that is, are merely concerned with subsuming phenomena under so-called covering laws, the *establishment* of these laws follows procedures which are describable and criticizable in value terms. They must be logically correct, avoid undue complexity but seek simplicity and elegance, and, above all, respect the conditions of relevance, i.e., contribute to the solution of problems of interest to the scientific community. It is in this way that science is one of the most valuable enterprises of humanity.

The Apex of Life

Let me review my story unfolded so far. When life came into being on earth I found out for the first time what it means for an entity to care about itself. To a sentient being things matter and events are important. It is not indifferent to its environment, because what goes on in the living being's vicinity can be either helpful or harmful, spell survival or demise. In a potentially hostile world the survival of an organism is a success story, an achievement of a good. Prior to the emergence of life I could not report such success stories, because there were in me no beings that cared, even about themselves. The electrons spin "happily" around their nuclei, no matter what happens. But an amoeba struck by a sharp protrusion loses its integrity as a being; it ceases to exist.

Thus, sentience introduced into me the primordial sense of value. Life is the carrier of what is valuable; to live is to seek the good of continued existence, to aim at success, realize states that are attractive, worth bringing about, admirable in themselves. All these alternative but conceptually related locutions help to set life apart from all other phenomena known to me up to this point of my career.

The concept of value stands for this entire spectrum of related concepts, and in developing my story further it will be useful to invoke other variants of this idea. Applicable to all life forms in at least partial ways, value takes on greater specificity of meaning in the human context. Unsurprisingly, in that context, anthropomorphism is inescapable but by no means regrettable. Even though full-fledged use of language arose only in the human context, it was developed by beings who wrought their language amid and surrounded by other forms of life. The meanings of such words as "pain," "pleasure," "fear," "anger," or "satisfaction" take into account not only human behavior but also the situations of creatures whose structures, functions and behavior are in many ways similar. Starting with reactions common to all living creatures, humans can also incorporate into language the differences among them. Hence, the anthropomorphic origin of language by no means entails scepticism about other, nonhuman forms of life. Although the meaning of "to intend" is taken primarily from the human context, what intention is can be learned from observing a cat stalking a bird, or what pain is from stepping on a cat's tail.

The panorama of my pre-human life was rich and spendthrift, as Annie Dillard happily put it. In the course of evolution I have explored and enjoyed immense varieties of sentience in countless species. I must admit, however, that the success of survival and my various degrees of wellbeing depended on fortuitous instinctive mechanisms devised across the eons of my organic evolution. My early forms of sentience entrusted themselves to species-instincts, each member of the species being almost totally dependent upon their automatic functioning. The welfare-downfall game was essentially out of the individual members' hands; the events in the drama of life were, so to speak, arranged for them in the structure and function of their organisms.

Only with the appearance of the *homo sapiens* did the situation change. And it changed radically. In place of, or at least in addition to, the instinct-governed, species-specific forms of behavior, I have hit upon a new method of determining and securing what is valuable, admirable in itself. That method is lodged in the ability to single out specific standards of what is good and to act in the light of these standards. The entire weight of value pursuit has shifted from the inchoately groping life-seeking molecules, through the instinct-governed species-determined form of sentience, through pre-linguistic socialization and enculturation, to the conscious identification and adoption of standards by the members of human groups. Humanity became the center of my cosmic drama; its role is to be chief protagonist in my pursuit of what is good, admirable in itself and worth pursuing for its own sake. The story of humanity needs to be now rounded out by noting one important fact: the actualization of humanity takes place in *persons*. Humanity is an abstraction; persons are concrete entities. It is in persons that my enthusiasm for life and value reaches its peak.

2

Person: My Favorite Mode of Being

❧ From Species to Individuality ❧

Self-Reference

Individuality is not in evidence at the level of my instinctive intelligence. An organism may perform a useful or even vital function for itself and for its species without being aware that it is doing so. A homing pigeon goes "home" by returning to its point of origin, but is not aware that *it* is contributing to some significant activity. It simply *follows* the instinct. The pigeon, or any other instinct-governed animal, does not know what it is to be *individuated*. Czeslaw Milosz has captured this unknowingness in three lines of a short poem ("What Does It Mean"):

> It does not know it glitters
> It does not know it flies
> It does not know it is this not that.

Individuality does not necessarily come into being with the advent of rationality, that is, with the ability to act in accordance with an acquired capacity or skill, such as speaking a language. The process of learning to act in accordance with a common standard is dominated by conformity to the *commonness* of the standard. The awareness on the part of the individual that he or she, by the very act of conformance, is establishing a common practice need not be present. Thus even on the human level, where rationality first comes into being, it is possible to act in unison with the group without being aware that in doing so the individual is acting in the name of or on behalf of that group. As Nietzsche observed, "we" is older than "I." Social cohesion or "solidarity" that emerged in the early stages of the human species, although it differed radically from the social cohesion of ants and bees, was nevertheless dominated by values securable by and for the group. Its members acted "as one," never differentiating themselves from others.

But one feature of language-governed behavior, in contrast to instinctual action, made the awareness of differentiation possible. That feature is *referentiality*. To refer is to pick out an object from other objects. There is only a short step from reference to *self*-reference. By speaking out a speaker distinguishes himself from other speakers. The speaker-hearer distinction already presupposes taking this step from reference to self-reference. I cannot overemphasize the importance of this change. From this point on my quest for values is no longer blind, pursued "through a glass, darkly," but is entrusted to self-conscious guidance of beings who, for the first time in life's history, see *themselves* as *agents*, as responsible for the enactment of standards of value.

With self-knowledge comes a new sense of power. If knowledge is power, so is self-knowledge. An organism capable of self-knowledge can realize that its behavior is not at the mercy of blind instinct, or at least that instinct is no longer its exclusive master. When self-conscious human beings cooperate, the master-slave relationship is transcended. As the standard-governed social life of a human group differs from the instinct-governed social life of ants, bees, schools of fish, flocks of birds, or herds of buffalo, so does an individual's self-conscious participation in common goals differ from thoughtlessly unhesitating conformism of a regimented society.

My path to self-consciousness, buried in dimly-recalled human history, was gradual, and I am not sure I can reconstruct it with any degree of confidence. It is possible that self-reference at first emerged as group reference, as "us" over against "them." Contrasts with neighboring human groups, in hostility or friendship, encouraged comparison or imitation, thus enabling human *groups* to refer to themselves, that is, to distinguish themselves from other groups and to acquire a sense of *social* identity. (It is of interest that such group differentiation often emphasized differences in language. For ancient Greeks their neighbors were barbarians, "the babbling ones," and for Poles Germans were "Niemcy"—the non-speakers, the dumb ones.) In this development common tasks and dangers contributed to the sense of cohesion and unity. Nevertheless, the self-referential capacity of language contained the possibility of extending the process of differentiation to *individuals*, resulting in the weakening of their total submersion in the ethos and will of the group. It would occur to someone that what is presented as general will is really what *the general* wills, thus raising the question of the justifiability of his authority. Even when the authority over the group is unquestioned, that authority may be seen as exerted over every*one* or some*one*.

Personhood

Thus language, in virtue of its capacity to represent human beings as individuals responsible for the enactment of standards, makes it possible for humanity to think of itself as a collection of *persons.* From that point on persons become the most advanced, most refined foci of my quest for value. It becomes their prerogative to consider whether the standards of behavior laid out by the group are *good* standards, worthy of adoption. In this connection I should point out a conceptual inadequacy of the etymological root of "person" in the Latin *persona.* The word originally referred to a mask worn by Greek actors on stage. The mask had an opening through *(per)* which words could be sounded. The etymological basis of this word does not make it a fitting model for describing the relationship between the human individual and the group, for it suggests that a person is merely a passive transmitter of group values. A person, however, is more than that. A person is a filter through which values pass in order that they may be evaluated and judged. This evaluation and judgment need not be negative or critical. Proposed by other persons as a candidate for a measure of what is valuable, it may be accepted as such, enacted, defended, celebrated. But to be a self-conscious individual is to be in principle capable of judging the value of a proposed standard.

Not to be forgotten is the fact that it is individual organisms, including human beings, that are concrete carriers of life. They are the *actual* instances in which I exist in my vital mode. On second thought, I must correct myself and retrospectively admit that the thrust toward personhood is not completely new at the human stage. The very beginnings of life in me were groping toward individuality. A minimal individuation is present at life's very start. Even a one-celled organism is an individual being, a *center* of feeling. Individuality and sentience are conceptually and not just empirically connected. A living thing senses its own differentness from its environment. In seeking to preserve itself, to remain intact and to maintain its bodily integrity, it is manifesting the thrust of life toward individuation. It expresses my desire to exist in the mode of feeling, sentience. On the pre-human level, however, that desire is not conscious. Personhood, on the other hand, is the most developed and fully conscious manifestation of my satisfaction with individuation. While *all* life exists at the level of individuals, the human person is its most individuated form. (Arthur Schopenhauer perceived the importance of this distinction in his exaggerated claim that all other animal species are individuals but that each human individual is a species in itself.)

No Group Mind

Humanity lives *in* individual persons. There is no such thing as a group mind. All of my feeling, thinking, judging, suffering, celebrating—in other words, all value realization—takes place in individual persons. The grasp of a meaning takes place in individual minds. Knowledge is always *some person's* knowledge. Understanding, appreciation, and enactment of a social practice or norm—intellectual, moral, or artistic—do not transpire either "outside" or "between" minds. To speak of a "collective mind" is to resort to a figure of speech, indicating perhaps that the content of what several or many persons think, believe, or cherish is shared by those persons. There are, of course, situations where many individuals do share the same perceptions or emotions. Members of a group watching a fire, an athletic event, a musical performance, can be presumed to undergo similar experiences. Nevertheless, in such situations the "group experience" is but an aggregation of individual experiences.

Of course, the fact and the consciousness of sharing an experience is an *additional* value that may be appreciated over and above the primary individual experience, whether it is listening to a musical performance, watching a dance, a touchdown, or a home run. It is important for the sense of solidarity of the group that its members participate in common life—practical, religious, or artistic, but even in order to be shared, the content of what is experienced must first of all be present in individual minds, reactions, and behavior. From the possibility of sharing experiences it does not follow that there is such a thing as a group mind. As there are no thoughts without someone to think them, so there are no values without persons to value them.

The temptation to postulate the possibility of a group mind may be rooted in the fact that practices, standards, and norms, and, more generally, all meanings and interpretations, are acquired *from* others as representatives of a cultural group. This temptation arises from the supposition that individual persons and their experiences cannot be the ultimate bearers of cultural values, because what they enact in their personal lives is *given* to them or bestowed on them by the tradition to which they belong. The tendency to think of persons as derivative entities, as merely *representing* their respective groups, becomes even stronger when, as we shall see, particular differentiations according to sex, race, or nationality are made to dominate personal self-consciousness. I believe there are reasons to resist this tendency, but of that later.

Personal Judgment

In reaching the level of personhood in the human race my adventure in the realm of value has reached its most intense and refined stage. No other form of life can realize values in such explicit and intense ways. Persons feel responsible for realizing group values when they not only act in the light of shared rules and norms but are also capable of *judging* whether some standards have been attained in their behavior. Success or failure are possible only when the *point* of standards is understood and appreciated. Only at the level of deliberate action can one speak of the presence of purpose in the full sense of the word. When one speaks metaphorically that nature has purposes and scores achievements, one borrows the notion of success from the realm of deliberate, standard-governed human action. No one faults nature when an acorn produces a mere shrub rather than an oak tree, or when a ewe throws a three-legged sheep. It is of course possible to look at nature *as if* it followed or failed to follow desirable standards. But this is only a metaphorical way of speaking because, as I admitted, at the state of pre-human nature I "obey" its rules blindly and unreflectively. As nature I don't make choices; I follow the instinctively developed patterns.

Personal judgment emanating from self-reference and self-consciousness is not a purely cerebral or intellectual affair. Like my other forms of life, persons are sentient beings, and no activity of a sentient being is devoid of feeling. Feeling accompanies every intellectual activity as well, even though that feeling may be characterized as "dispassionate" or "disinterested." Doing mathematics, for instance, involves, like any other human doing, not only the person's brain but also the whole body. A person engaged in solving a problem is usually aware of that for the sake of which the activity goes on. To proclaim the human capacity for thinking self-contained is to forget that thinking is taking place in a sentient creature who experiences that thinking in multiple ways. The active character of thinking is especially evident in situations when the thinking or reasoning is not routine but enlists special attention and effort. Even though the tension of that effort may be first experienced as unpleasant, it gives way to feelings of satisfaction and occasionally of exhilaration when a difficult problem is solved. Every thinker, great and small, knows a "Eureka" experience.

Persons: Bearers of Culture

The concomitance of thought and feeling is present in all human activities. Expertise in mathematics or physics no less than in music or dance gives varied satisfactions to the experts and to those who benefit from their expertise. The activity of transmitting knowledge from person to person, from generation to generation, engages not only people's minds but also their emotions. Places of learning, such as modern universities and their ancient predecessors, are prized as indispensable to the life of a civilization. But the transmission of knowledge is always *through individuals*, and such a transmission is more than just intellectual process. No civilization is thinkable without persons whose involvement in it is not only representative but also creative. Neither the rise nor the fall of civilizations can be adequately accounted for without reference to individuals who, positively or negatively, played central roles in their people's history. Without intense individual mastery of specific forms of life characterizing a culture, that culture cannot continue. When the transmission of that mastery ceases, a culture exists, if at all, only in memories, books, and libraries.

Some good examples of the crucial importance of individual mastery for the ongoing social enterprises can be taken from contemporary civilization. Consider an aviation technician inspecting a large airliner between flights. Even the routine skills he applies in his job have to be applied with attention and concentration because they are vital for the people using the aircraft. Notice how many complex skills (some of them, those of the pilot, for instance, are highly sophisticated, requiring years of training) have to be carefully coordinated if the plane is literally to get off the ground. Or consider an astronaut guiding a spaceship around the earth's orbit or to the moon's surface. Countless persons have poured their respective masteries into every single item and function of the ship. It can perform its awesome task only because the most advanced knowledge and creative intelligence of many generations of persons contributed to this success.

The flight of the spaceship is but one example of a *deliberately* created nexus of individual performances. Any noteworthy enterprise of a culture rests on such performances rendered by those who understand the intricacies of the enterprise, be it primarily technological or scientific, or religious, political, or artistic. A scientific laboratory, a museum, a modern hospital, an architect's studio, a legislative chamber, a church service, a university seminar—all those contexts display activities whose collective meanings depend on individual contributions by persons wholeheartedly participating in these activities. In all of them I want to call attention to the crucial function of individuals who act as originators and developers of collective cultural achievements.

The role of the individual in envisaging, creating, and maintaining cultural institutions, structures, patterns, and practices has not gone unrecognized. Not only individual observers but also entire schools of thought, philosophical and religious, have proclaimed the importance of human individuality. Descartes noted in his *Discourse on Method* that every important achievement is primarily due to the solitary effort of one person. To illustrate this claim he observed that a city designed by a committee is bound to suffer in comparison with that designed by a single individual. More recently, John Stuart Mill, in a ringing manifesto *On Liberty,* defended individualism on the grounds that it alone forms the basis of a healthy society. Another nineteenth century thinker, Friedrich Nietzsche, was of the opinion that Western civilization, in spite of its lip service to the importance of the individual, still has not begun to recognize that the course of human history turns not on impersonal social forces but on the efforts of outstanding creative individuals. Some of Nietzsche's epigrammatic and dramatic statements of this view lend themselves to unfortunate elitist interpretations, but when more soberly stated, his position is faithful to the mainstream of the Western civilization, in which the special status and dignity of persons is proclaimed and celebrated. Christianity speaks of the infinite worth of every person, and Kant's moral theory, undoubtedly the most influential theory of modern times, places the burden of rational and moral action on the will and intention of the individual. In the entire Western tradition the dignity, worth, and autonomy of persons is made central.

Tasks of Personhood

The changes which are brought about in me through the efforts of humanity embody purposes that are envisaged and articulated by persons in standards of action they set for themselves. This capacity shows forth most fully the distinction broached previously between natural and human purposes. Only the latter show themselves in explicit *tasks*. A task differs from a natural purpose precisely because of its origin in a verbally formulatable design, emerging in a person's mind and communicated to others. What is communicated in language is not just the formula but also the motivation to organize human behavior, attitudes, and feelings around a specific task. That task may be practical or theoretical—building a bridge, designing a scientific experiment—or it may aim at a change in a personal or social dimension—proposing a marriage, or setting in motion a political program. Human history, on the personal and social levels, in private and public life, is the story of successful or unsuccessful enactment of such tasks, each manifesting a particular instance of my insatiable quest for continually refined and now consciously and deliberately pursued values.

Human story on earth, along with thousands of failures, destructions, and crimes, has also produced, through genius, invention, and cumulative sharing of knowledge and inspiration, worthy monuments to humankind, thereby testifying to spectacular successes of life. Armed with language and thought, life now has unprecedented options: through persons it can *determine* what is good, is worth pursuing. The essence of human freedom resides in the ability to aim deliberately and self-consciously at the production of affairs which now, for the first time in my history, can be *judged* to be good in themselves.

In the human form of life individual actions are guardians of value. As such, they are radically distinct from instinctive behavior. The latter is governed by pre-existing structures, and its importance is not evident to the organism. If one could speak here of representation, instinct-governed sentient beings do not represent themselves but their whole species. (Descartes went overboard, however, when he, failing to see conscious agency in animals, deemed them to be mere automata. On the other hand, Schopenhauer was exaggerating when he said that each human individual is a species of its own.) While instinctive behavior functions on behalf of the species, human action occurs on behalf of consciously identified and chosen standards, in the light of which the decision is made of what is to be regarded as good.

The conclusion I wish to draw is that in humanity I can, for the first time in my history, exist in the personal mode. As I argued, even the social dimension of human existence is experienced on the individual level. If humanity is the latest and most advanced form of life, persons are *spokespersons* for humanity. In its unique way each life actualizes what is possible for humanity and hence is responsible for the ways in which the human form of life comes to expression.

❧ From Nature to Morality ❧

Caring

The earliest lesson I learned when my matter discovered the possibility of existing in a living form was the importance of *caring*. Feeling and sentience are primordial phenomena of caring. They arise when something in its surrounding matters to a living being. Since the very definition of sentience relates the organism to its environment, caring acknowledged the *objectivity* of environment. The feeling is subjective but that which gives rise to it is not. Feeling is a response to an objective situation. Recall also that the discovery of feeling is tantamount to the discovery of values. Thus from the very start my awareness of values arose in objective contexts; it never is something merely inner or subjective but involves an interaction between an organism and its environment.

When, with the proliferation of life, living organisms needed to react to other organisms of the same species and to other forms of life, the phenomenon of caring had to define itself in new ways. I have already referred to symbiosis, the tendency of organisms to develop an accommodation, a *modus vivendi* with other organisms. Human beings are now coming to recognize that all of nature is an ecological system, that various forms of life need and support each other in their *conatus.* Although it would be an exaggeration to say that life *respects* life, it is nevertheless true that in fact the actual display of life on earth follows the rule: live and let live. I disagree with those who regard aggressiveness as the primary characteristic of life. Life is aggressive only when it has to be. A lion hunts only when he is hungry; at all other times he is content to be absorbed in the enjoyment of living processes within and around him. Lacking language he has no means of attributing qualities to other living beings—plants or animals—but he senses their presence and can tell the difference between organic and inorganic matter and between living and non-living things.

Instinctive mechanisms developed by various species in the course of evolution are adjustments in behavior which take into account both the physical environment *and* the presence of other forms of life in it. Only in this way could symbiosis occur. Indeed, symbiosis is nothing more than the fact of such adjustments of organisms to their mutual involvement in an environment they have to share in order to survive. The marvelous ecological interdependence of the earth is but another witness to the fundamental sense of caring which life in me manifested from the very start.

That mutual adjustment and harmony are a function of caring couldn't dawn on me of course until, in the human form of life, I discovered reference and self-reference. That discovery in turn made possible the very idea of *responsibility.* That idea could not arise until an action could be ascribed or self-ascribed *to* someone; that is, until identification of an agent, by reference, could be brought off. The context within which this was done explicitly is the human social context. But even human beings exhibit two different kinds of cooperation: one in which the will and coercion of the group is dominant and one in which the participation of the individual is deliberate and voluntary. The distinction is extremely important. In fact, it gave rise to an entirely new phenomenon of life: *morality.*

Morality

Philosophers still spend a great deal of time trying to give a convincing account of morality. Many a theory has been propounded, and most of them have something valid to contribute. It is useful, however, to point out features that characterize a *transition* to morality from a merely natural behavior. In natural behavior the contribution of the individual need not be taken into account to describe what is happening.

The transition has already been made when a person asks the question of which philosophers seem to be especially fond, "Why should I be moral?" The question is problematic because it seems to presuppose that both the questioner and the questioned understand what the word "moral" means. So to deal with the question at all, the meaning of "moral" must be explained. One can begin by noting that the question also contains the word "should." What is the meaning of *that* word? A person who uses it in the sentence "Why should I do X?" may simply mean "What advantage do I get from doing X?" or "What is there in it for me?" If this is what he means by "should," he does not yet understand what it is to be moral. He does begin to understand it when he realizes that morality comes into being when a person can *take an interest in the interests of others.*

This step is a fundamental one and is not reducible to any interest of one's own. It is taken when a person no longer is merely coerced by the will of the group but treats the needs, wants, and desires of others as objective values worth honoring for their own sake. It is in this way that a person becomes moral, that is, capable of freely and deliberately acting on behalf of others. Another way of putting this point is that a moral person becomes an active supporter of the value of a *standard* and treats that standard as objectively valid for all members of the group. The recognition of the validity of that standard translates into a sense of responsibility *for* it.

The importance of this step cannot be sufficiently stressed. Once it is taken, the process of bringing values into the world is drastically transformed. By encompassing moral concerns, that is, acting in the interest of shared values, rationality becomes self-conscious; its upholders see themselves as responsible for the nature and character of standards governing the life of the group. One should be careful to note that the use of the word "governing" here applies not to some impersonal social force hovering over the group and forcing its members to "toe the line." It refers to the fact that the application of standards is maintained by a free, self-responsible commitment of each member of the group to shared standards of morality. The governance of morality is the governance by moral, mutually supporting individuals.

A transition occurs which can be described as a transition from nature to morality because the pursuit of values prior to moral action is the work of nature alone. Natural behavior includes the general *conatus* of life, instinctive mechanisms, and group coercion or herd-instinct, if you please. Even in the latter the role of the individual is completely passive because at that level group dominance is unquestioned. "Herd-morality" is a misnomer, a contradiction in terms. The behavior of a herd is natural—neither moral nor immoral, because morality exists only when standards of morality are filtered through the judgment and commitment of individuals. One might say, then, that morality is "artificial," a product of human art or contrivance; it is not an automatic, mindless biological process.

Prudence

Some far-reaching consequences follow from my discovery of morality in human beings. Their full range needs to be laid out in some detail, but before turning to them further elucidation of the nature of morality should be made in order to avoid a possible misunderstanding. One such misunderstanding is the confusion of morality with prudence. Since rationality, in contrast to instinct as merely species-intelligence, is an ability to act in the light of some standards, prudence is certainly a form of rational behavior. It embodies the principle of subordinating one's immediate good to the pursuit of a long-range and more inclusive good. But since it is limited to self-interest, the rationality of prudence has not quite reached the level of a rationality which is also moral. Nevertheless, prudence is at least a conceptual cousin of morality precisely because it lies halfway between nature and morality. The fact that self-interest can be long-range calls into being dispositions that are conducive to morality, namely the suppression of momentary impulses for the sake of a larger good. This feature, by encouraging rational self-control, makes it possible to thwart merely egoistic, self-concerned impulses and may result in recognizing the objective validity of the interests of others. Thus prudence foreshadows at least the possibility of altruism as a basic moral disposition.

Although prudence may be seen as a transitional step toward morality, genuine morality requires an additional step: treating legitimate interests of others as being good in themselves. There is no way to avoid the collapse of morality into a truncated or phony form of itself except by insisting that it calls for taking an interest in the concerns of another without reservations and for their own sake. Unless one can see and appreciate the fact that the pursuit of the interests of others is objectively good in itself, morality cannot and does not get started. Its starting point is genuine only when one can honestly say "You are entitled to pursue your own interests. The world is the better for it, and I would not want it any other way."

Other Quasi-Moralities

Another misunderstanding of morality misrepresents the role of the individual in it. To say that individuals are responsible for upholding standards of morality is not to say that they are *arbitrary* legislators of it. In his famous and fundamentally correct articulation of what morality consists in, Immanuel Kant described it as acting in the light of the Categorical Imperative: "Act only according to that maxim by which you can at the same time will that it should become a universal law." Kant also claimed that the source of the Categorical Imperative is the rational nature of human beings themselves and not some external authority beyond humanity. One may be inclined to interpret Kant's view as an invitation to subjective willfulness, as granting the individual a right to a unilateral exercise of power or authority. According to such an interpretation, the person invoking a moral standard claims to have a sole, arbitrary jurisdiction over it. Such a person, in acting morally, is presumed to have the thought: "I, by this act of my will or commitment, grant you the right to pursue an interest of your own."

To conceive of the role of the individual in this way, is to misunderstand Kant and to nullify the condition specified before, namely, that a moral individual acknowledges the objective validity of the value of interests pursued by others. This is what the Categorical Imperative calls for. Morality, in other words, is another, consciously articulated, expression of *caring*. Morality, like symbiosis, expands the caring capacity beyond the confines of one's own skin, so to speak. To deny that it is often actual is the same as denying that the symbiosis of pre-human organisms is actual, that a lioness does not nurture her cubs. Morality is no more, and no less mysterious, than the tendency of nature to live and to let live.

To realize this is to steer clear of another mistake regarding the nature of morality. A moral commitment is sometimes interpreted as an initiation of a kind of bargaining. "I grant you this right because I expect you to grant me my right to pursue similar interests." Such a formulation undermines morality right away. For if the real reason why a man grants rights to others is because he wants them to respect his rights, and if he respects their rights only on *that* condition, the "morality" in question collapses into self-interest, or at most into prudence, which, as we have seen, is but a gateway to morality.

Demandingness of Morality

Although morality as concern with interests of others at first may express itself at the level of refraining from harming others and of not interfering with their comings and goings, it also calls for moving beyond that minimum requirement. Indeed, morality in its more sturdy varieties often requires a painful effort, even sacrifice. Moral ideals are often challenging and difficult to live up to. That is the reason why morality is sometimes represented in negative terms; it limits options, constrains wills, frustrates desires. It may even appear oppressive and joy-killing. But this situation only shows how far-ranging moral sensitivity as a sophisticated form of caring can be. To capture its scope civilized societies have invented a large vocabulary describing some special phenomena of moral life: conscience, compassion, guilt, regret, atonement. The more demanding forms cf morality presuppose an ability to take the first step already identified: the willingness to include the fulfillment of the interests of others among the events in the world that are objectively worth bringing about by one's actions.

It is not surprising that morality should appear in a negative light; the tasks it imposes are difficult because they sometimes interfere with satisfactions persons ordinarily seek. Morality demands that they do the right thing, regardless of whether it will give them pleasure, and they know in many instances that it will cause them pain. Kant claimed that the opposition to natural inclinations, which are usually self-serving, precisely characterizes a moral situation. Morality, he says, strikes down natural inclinations and has the authority to overrule them. And yet, he adds, the very act of subduing these desires produces a special kind of feeling, the feeling of self-respect. It is experienced when a person finds it right to pursue ends that are good in themselves even though they do not result in the satisfaction of one's own desires. Such a person is moral precisely because among his desires there is also a desire to *have* moral inclinations, that is, dispositions to respect the needs and interests of others. It is important for such persons to view themselves, and be viewed by others, as capable of acting on principle, which enjoins them to forego the satisfaction of a desire when that satisfaction is secured at the expense of someone's discomfort or suffering.

Kant was mistaken in thinking that morality must always conflict with natural desires. Any genuinely shared enterprise is a counter-example to this supposition. For example, think of the satisfaction of a performer in the chamber music quartet. By disciplining himself to play his part well, he finds his own satisfaction, but he also contributes to the satisfaction in the performance on the part of the other three players. (This point holds even in cases where there is no audience, whose enjoyment would, of course, be an added, yet different satisfaction.) In order to make the experience of others a success, objectively good, a player feels obligated to play his part well; he takes an interest in his partners' interest to hear a good rendition. But in this case, playing and hearing such a rendition is also *his* interest and gives *him* satisfaction.

Although this example about a joint venture such as a musical performance is not explicitly moral, it fits moral situations as well. It parallels any instance of sociality or of solidarity with a joint enterprise where participants are aware of the value of the activity, both for themselves and for others. They can derive satisfactions from the cohesiveness and mutual supportiveness among members of a family, or from multidimensional participation in the affairs of the whole community. In its deeper reaches morality brings into being a genuine sense of mutual acceptance; it fosters the sentiment of generosity, in which cooperation is not seen as merely demanded or expected but as satisfying the participants' own minds and hearts.

Evil and Crime

If the essential step toward morality is caring, taking an interest in the interests of others, then there is no difficulty in understanding evil. Where the capacity to care is absent, evil actions can be expected. Philosophers rightly distinguish between natural and moral evils because behind harmful effects of natural events there is no *intent* to harm, in contrast to actions of human beings. Understandably, it is the moral evil that is the focus of my attention because in contrast to natural events human actions are in principle controllable by the capacity to judge. Evil persons either fail to respect the interests of others or they ignore them. For the sake of some real or imagined personal satisfaction they deliberately and gratuitously inflict pain on other living beings.

Aware of the total sweep of life on earth, I would be the last to deny that moral evil occurs. Many people perform evil actions and, even worse, in virtue of performing such actions habitually and routinely become evil persons. But why is the existence of evil problematic? It is because in reaching the human stage I have come to understand and to appreciate the special value of morality. For one thing, values securable by moral behavior are of great importance for the optimal functioning of personhood, which in turn produces productive, creative, and happy societies. That is why societies in which moral evil takes hold are frustrated and paralyzed with fear. When evil becomes frequent and flagrant, giving rise to endemic crime, the very possibility of continuing civilized existence may be put on the line.

There are those who connect evil and criminality with aggressive animal behavior manifested by life in its "state of nature." The state of nature, claimed Thomas Hobbes, is characterized by war of each against all, and life in that state is "solitary, nasty, brutish, and short." It is no secret that in my pre-rational *conatus*, in which self-preservation is the supreme law, cruelty of one species toward other species, or a fierce competition for food or mates among members of the same species, is the order of the day. But since my main objective in all forms of life is to live and not to exterminate, aggressive behavior never is gratuitous but is limited to self-preservation.

Cruelty, however, was only a function of necessity and never belonged to my favorite values. That is the reason why I have pressed on toward a species, the human species, in which the problem of survival could be maintained by cooperative behavior, thus minimizing cruelty and introducing morality. While I do not deny that evil motivates many representatives of the human species, I consider it a vestige of my more primitive form of life. Having made a transition from nature to morality, I set myself squarely against all forms of evil. In that effort I find my voice and my will in persons and societies that champion the superior value of morality.

The task is not easy. The possibility of lapsing into evil, spurred on by crude and raw onslaughts of *conatus*, often crops up from below the surface of life, requiring of morality constant vigilance. Although wholly evil persons are rare, every person may occasionally succumb to evil. Some circumstances aggravate this danger. When, for example, people's mode of life develops powerful dependencies and addictions—transforming their organisms into insatiable helpless craving mechanisms—the susceptibility to the needs and interests of others is weakened and may be wholly destroyed. The result is an absence of even elementary moral dispositions and a concomitant tendency to do evil.

Another context in which insensitivity to the horrors of evil may be fostered is political. By portraying one's ideological enemies as harmful, parasitic, dangerous, or even "inhuman," a political program can condition its adherents to behave with extreme cruelty toward other human beings. Bitter and prolonged wars provide a likely context in which cruel and brutal tendencies are elicited. Other contexts in which evil can flourish are found in fanaticism, fostered by relentless indoctrination and brain-washing. Caught up in such contexts persons lose sight of the dividing line between good and evil, decency and decadence. Or at least that line is easier to cross, exposing human hearts to the bewilderment of immorality. Alexander Solzhenitsyn, forced by the political circumstances of his time and place into such a context (labeled by him *The Gulag Archipelago*), gave the following account of the human vulnerability to evil:

If only it were all so simple! If only there were evil people somewhere insidiously committing evil deeds, and it were necessary only to separate them from the rest of us and destroy them. But the line dividing good and evil cuts through the heart of every human being. And who is willing to destroy a piece of his own heart?

During the life of any heart this line keeps changing place; sometimes it is squeezed one way by exuberant evil and sometimes it shifts to allow enough space for good to flourish. One and the same human being is, at various ages, under various circumstances, a totally different human being. At times he is close to being a devil, at times to sainthood. But his name doesn't change, and to that name we ascribe the whole lot, good and evil.[3]

A lesson to be learned is that some circumstances and some arrangements increase the possibility of allowing evil to flourish. Therefore, an important task of morality is to avoid such circumstances and arrangements. Knowing that under some conditions moral tendencies will be either prevented from developing or crowded out—either by the loud demands of the addicted body or by the distorted perceptions of the brain-washed psyche—a wise society will not allow such conditions to obtain.

❧ From Integrity to Autonomy ❧

Moral Integrity as Achievement

Morality, I have claimed, is my step *beyond* nature. That step involves the capacity on the part of human beings to escape the natural constraints of instinct or of blind conformity to societal norms. I don't want to suggest that morality is *un*natural. In a broader sense of "nature," as the totality of all energies coming to expression in my cosmic career, it is *natural* for human beings to be moral. Morality is an actual phenomenon, clearly observable in histories of persons and societies. It is also natural in the normative sense; humanity has found its own advance toward morality a *desirable* development. Morality makes it possible for new kinds of values to emerge, values that stem from voluntary and deliberate cooperation for the sake of ends which can be realized only through free commitment to advance the interests of others.

One of the values to which morality gives rise on the individual level is integrity. Persons of integrity are reliable; their actions manifest a steady commitment to ethical principles. It is not surprising that honesty and integrity are often mentioned in the same breath. To have integrity is to be trustworthy, to respect the standards of truthfulness. Conversely, to lack integrity is to be unreliable, fickle, disloyal. The mark of integrity is steadiness, that of its opposite, changeability. With these characteristics go corresponding pictures of the nature of persons exhibiting them. To acquire integrity is to endow one's character with a certain weight, solidity, substance; it is to persist in certain definite attitudes. Conversely, to lack integrity is to be a "lightweight," easily swayed by circumstances and thus to be unpredictable, flighty, scattered, or chameleon-like.

Unlike instinctive or herd-like behavior, the behavior rooted in moral integrity does not just happen; it is an achievement. Integrity must be valued by the person exhibiting it; it is the person's voluntarily and deliberately maintained disposition. It cannot be sustained by external pressure alone. It is not a function of social environment, even though social conditioning does play a role in the emergence of personal integrity. The necessary condition for the presence and survival of moral integrity is the person's own conviction that its maintenance is desirable, that it is valuable and admirable in itself.

Morality, however, does not exhaust the range of possible values which persons can pursue. Indeed, an exclusive preoccupation with the moral side of life may be limiting. (Arthur Schopenhauer described Fichte, a fellow-philosopher who saw morality as the end-all and be-all of life, as a "moral windbag.") It is not difficult to see why. Moral standards specify conditions under which persons can pursue their interests without fearing that this pursuit will be interfered with and in the assurance that other members of the community welcome that pursuit as desirable. Thus morality presupposes the antecedent conviction that personal flourishing is objectively valuable. It secures certain humanly desirable goals. While morality itself is one of the objective values to be pursued, it is not the only one. Moral integrity is an end in itself, but it leaves room for activities that are also intrinsically good.

Autonomy as Self-Development

This is the place in which it will be helpful to introduce the notion of *autonomy*. Autonomy includes moral integrity but also goes beyond it. Morality is taken seriously not only because it is good in itself but also because it helps persons to bring about the realization of their individual self-conception. Besides being committed to some moral standards, persons face the task of integrating these standards into their overall plan of life. Autonomy is the capacity to strive toward an inclusive unity and harmony as characterizing a person's journey through life. Autonomous persons work seriously on formulating their goals and objectives and seek to establish a personal manner in which they would like to pursue them. Notice that the motivation behind my thrust toward personal autonomy in individual human beings is the same as that which freed them of their dependence on absolute authority of the group. In either case the result is a refinement, an enhancement of my quest for value. In the human form of life I have realized that values achievable through morality and autonomy are superior to all others I could realize up to that point in my career. It is why they are at the center of humanity's attention.

In aiming at autonomy persons try to harmonize their traits, capacities, and dispositions with outer circumstances and objective conditions of their lives. This task of unifying experience into a coherent whole is a life-long task. Like all other life phenomena, it takes place in time. Everything that persons do involves the integration of present, past, and future. Present human action arises from the past and points toward the future. All action involves memory and anticipation. To understand what a person is doing is to know the context from which the action springs and what state of affairs it aims at. To the extent that such knowledge is not available either to the agent or to the onlookers, the action is problematic, and in limiting cases it is quite legitimate to question whether there is any *action* at all. Neither the agent nor the onlookers understand what is going on.

The Scope of Autonomy

The integration of present, past, and future in a person's actions can be either momentary and superficial or long-range and deep. The former need have little or no connection with the person's autonomy; the latter exhibits some central features of that person. Actions which reflect a person's long-standing goals and commitments reach deep into the past and project far into his or her future. The more an action expresses such deep commitments and lasting projections, the more characteristic it is of the agent. A person who always lives from such deep and authentic resources would be able to declare: Every moment of my life is the whole of my life. Such a declaration, of course, is more plausible as a statement of an ideal than of an actuality. As an ideal it expresses the desire to achieve fullest autonomy: utmost collectedness and directedness in every single thing one does. Every action would arise out of what one really is and aims at, or what one would really like to be. Correspondingly, some actions would be seen as not at all characteristic of what one is, or would be condemned as wasteful of one's powers.

Granted, there is something frightening about such a concentrated and self-consciously directed lifestyle. In extremely single-minded individuals one may rightly suspect rigidity or fanaticism. On the other hand, an internally concentrated lifestyle may characterize a genius. Misgivings arise only from the suspicion that an intense single-hearted devotion to a life plan is likely to narrow a person's vision of desirable possibilities. People rightly worry about child prodigies becoming freaks, or asocial and emotionally starved neurotics, whose development in many areas of human excellence is arrested. But there is nothing in the notion of autonomous personality that necessitates narrow-mindedness or stunted growth. On the contrary, a mode of life which overlooks some key values that are worth including in the scope of one's interests is to that extent deficient. Because a creature with a mind is in principle capable of surveying an entire horizon of desirable possibilities, in the quest of autonomy the matter of scope should not be ignored. Autonomy bought at the price of ignorance about some typically desirable excellences is of questionable value. Such losses and limitations may be excusably overlooked in a life of a genius whose concentration on a narrow area of competence produces spectacular results. But even in such cases, one wishes that that life were capable of incorporating other humanly desirable traits as well. A genius need not be a crank.

Apart from special cases, the search for autonomy combines many objectives. Most persons are equipped with dispositions, talents, and capacities that do not run on a single track but are multidimensional. Taking account of the value of each dimension, each to be acknowledged in its own way, at the right time, at the appropriate junctions of one's life, persons can act in ways that reflect this many-sided search for a comprehensive autonomy. There is more than one way of being oneself, more than one manner of expressing a side of personality valued for its own sake. But the question worth asking repeatedly is whether the many strands which an individual wants to weave into the fabric of his or her life do amount to a harmonious whole, and whether they add up to the kind of person one wants to be.

Such a search, of course, cannot be pursued without a risk of error and transgression even against one's own ideal. Neither the knowledge of external facts nor the self-knowledge of a person are always equal to the actual tasks and tests of life. Experience *is* the best teacher; but to be teachable one must be *open* to experience, with all its pitfalls and dangers. It is also possible to overreach oneself; well-roundedness is easy only if the person's radius is short, and broad-mindedness may degenerate into flat-headedness. Throughout the entire life-long journey human beings need to keep an eye on what kind of person in them is in the making. The ultimate aim is to integrate all constructive possibilities, inner and outer, into a whole that is truly in harmony with the person's self-conception. The resulting autonomy is a paramount value which I have discovered in life.

Imbued with the desire to attain autonomy, which I believe is desired by every human being, persons may be able to say: this is a good life, good in itself, worth affirming and celebrating. They can say this if their individual actions fit their overall self-conception, realizing their strengths and avoiding their weaknesses. To the extent that persons' lives and actions are in their own hands, they are entitled to claim autonomy. Of course, self-deception, rationalization, and unjustified self-praise are not something that can be avoided altogether, but those who deny themselves the right and the capacity for a more or less realistic self-appraisal fall short of what is possible for them as persons.

Obstacles to Autonomy: Elusiveness of the Self

The task of achieving an integrated, autonomous personality is not easy; the path toward it is strewn with obstacles and perils. One such obstacle, surprisingly perhaps, lies in the very nature of thinking, the capacity made possible by language-governed reference and self-reference. Human thought can fall prey to what could be called the seductiveness of reflection. When one thinks of something, one can be easily seduced by a neighboring thought coming by way of association, or contrast, or denial. Allowing the stream of consciousness to flow on, a mind dissipates its attention. Of course, thoughts need not wander aimlessly. Since the objective of thinking usually is to survey the existing possibilities for their interest and relevance, the inherent restlessness of thought is not undesirable. It becomes so, however, when there are good reasons to stop further explorations and to dwell on some experiences in order to savor and digest them, to absorb what they have to offer and relate them to the course of a person's experience.

Centeredness of attention is possible, however, only when one assumes that there *is* a center, when one can speak of a *self* that has experiences. Therefore, it is not just a parenthetical interest that the concern with the slipperiness and elusiveness of the self may undermine the very idea of there being a self at all, not to speak of an autonomous self. A philosopher who dealt explicitly with this problem was David Hume, and he drew a disconcerting conclusion from the central views of his predecessors, Locke and Berkeley. Hume concluded that the notion of the self is suspect because while examining the ideas one actually has at any given moment one finds that they include various perceptions and sensations, but the idea of the self cannot be found among them. Hence, Hume declared that human minds or selves are no more than "bundles of perceptions." Jean-Paul Sartre also found it impossible to pin down the self to some stable structure and coined such slogans as "man has no nature," or "existence precedes essence." For Sartre and his existentialist followers, this discovery prompted an exhilarated sense of boundless, absolute freedom. Originating on the sidewalk cafes of Paris, after the publication of Jean-Paul Sartre's *Being and Nothingness* in 1943, this mood affected the intellectual climate in many parts of the world and found expression in such phenomena as the theatre of the absurd and even America's beatniks and flower children.

The restless, self-reflective character of human consciousness suggested to Sartre that there is something paradoxical about it: consciousness is what it is not and is not what it is. To think about anything is to be already beyond it. Because every thought is *about* something, it cannot coincide with its object; it runs ahead of it (or is "about" in the sense of being somewhere in the vicinity). As an act of consciousness, the thought is in a sense "detached" from its object and can carry in its wake other thoughts, thus increasingly, and paradoxically, taking a person's attention *away* from the object. Even while realizing that something is the case the mind is open to the possibility that it may *not* be the case. Every affirmation of a fact contains the negation of this fact as a logical possibility. But even a negation can take many forms; a thing or a state of affairs could be in various ways other than it is. To have any knowledge at all is to be aware of alternative situations that can be compared or contrasted with what one actually finds to be the case. Thus, endless dialectic takes over, keeping the person's mind unfocused.

The danger of undercutting autonomy, latent in the very possibility of thought and reflection, is reinforced by the fact that persons are not *just* thinking beings. If they were purely logical, intellectual creatures, perhaps their thoughts would be confined to a radius limited by logical or linguistic possibilities. But human thinking is also at the service of needs, interests, and desires. Consequently, the shifts in consciousness follow the multiple paths of associations conditioned by these needs and interests. Not being disembodied, thought is affected by feelings, wishes, and volitions. Buffeted by fluctuating moods, persons often fail to map a clear course toward chosen objectives. Often at odds with their present, tentative, or even firm decisions, their train of thought gets derailed, leaving them stranded among unforeseen possibilities.

A related type of obstacle on the road toward autonomy and self-control is *distraction*, a familiar phenomenon. It is easy to catch oneself thinking that whatever one is doing at the moment is inferior to what one might be doing, that things are more exciting elsewhere. It is extremely difficult to escape the feeling that the action is somewhere else, and the real show on some other stage. This stage is imagined to be either in some other part of space or taking place at some other time, either in the future or in the past, depending on whether a person is seized by anticipation or by nostalgia.

People should become clear as to what precisely is deplorable about this tendency. Although they can escape neither their past nor their future, they are well advised to give proper weight to events *when* they happen. Not to do so is irretrievable loss. Distraction deprives persons of the opportunities to affirm, appreciate, enjoy, and celebrate the good things that happen to them. Living either in retrospect or in prospect is only half-living. Since the sum total of my values is contained in the experience of all sentient beings, and since self-conscious experience of what is valuable takes place in autonomous persons, every failure to recognize and to affirm *occurring* values is a loss to me. My vital pulse slips by unrecognized and uncelebrated. How many times the value of a golden moment revisits many a person only as an afterglow! One realizes, regretfully, that the full-blooded reality of what one was undergoing passed one by while one's mind was elsewhere.

To live fully and abundantly, one should cultivate collectedness and centeredness. To neglect your present experience in favor of something going on elsewhere, either in someone else or in yourself at some other time, is to prefer shadow to substance. Such a neglect, if persistent, is bound to result in self-depreciation and self-alienation. Moreover, the slighting of oneself is likely to bring in its wake the slighting of the world as well. Among the consequences may be depression, pessimism, and cynicism.

Every actuality eventually leaves the stage of the present and recedes into the past. The question is whether *before* becoming a dead page in the annals of recorded or unrecorded history, the present is understood, recognized, appreciated, absorbed, and celebrated for its actual significance and value. Only by being awake and alert can one catch living experience on the wing, can present the potential values of the moment from going by ignored or unrecognized. Every person, in the actual present, is the center in which my potential significance is lived out. This is the deeper meaning of the frequently misunderstood, because narrowly conceived, motto: *Carpe diem,* seize the day. As Helen Keller wrote, "I who am blind can give one hint to those who see—one admonition to those who would make full use of the gift of sight: Use your eyes as if tomorrow you would be stricken blind."[4]

Obstacles to Autonomy: Scepticism of the Psyche

The natural desire to be an autonomous and self-reliant person has another potential enemy. Some popular theories about the nature of the human self encourage the conclusion that it is a nervous wreck from the very start. Freud's general theory of the human psyche is often understood in this way. Without denying the reality of repression and of so-called unconscious motives to which Freud has called attention, one may nevertheless ask some questions about them. Does the occurrence of such phenomena, which no doubt can do a great damage to human psyches, warrant the crippling conclusion that persons' control over their conscious lives is always suspect? Such a conclusion, alas, is too often drawn by those who tend to give primacy to the Freudian theory of the unconscious. If the Id is purely instinctive, and the Superego arbitrarily imposed on a person by parental and social authority, then there is little that the poor helpless Ego can do to get control over that person's life. (There is even the prior question whether the Freudian ego, being but a *fragment* of a person, has the ability to *act* on its own. *Persons* act, not egos.) The influence of the Freudian and post-Freudian psychologies has been so pervasive as to raise a serious doubt whether people can ever trust themselves in their self-appraisals. This scepticism encourages the tendency to take what anyone ever says with a grain of salt. Conditioned by this "scientific" view, the most that one is prepared to grant is that there are games that people play, or that they present their selves in ways that most of the time are self-serving, deceiving, or self-deceiving.

It should not be surprising that, being wedded to such a view, people have difficulty in trusting either themselves or others. The first, and often last, response is to doubt. Nothing, people tell themselves, can be taken at its face value. No utterance, one's own included, really means what it says— the real truth is behind it, hidden somewhere in the labyrinthine maze of ulterior or buried motives. In an age of scientific expertise, this theory-induced distrust soon pervades personal and social lives.

I do not deny that deception and self-deception occur. What is significant, however, is that if appearances always lie, then the exception has become the norm, and the norm the exception. If the human psyche is inherently untrustworthy, then it is impossible for people to arrive at a confident judgment about their own integrity, and any expression of confidence about being able to arrive at a true appraisal of one's aims and objectives is immediately suspect. As long as this scepticism keeps its hold on people, it will not be possible for anyone even to *attempt* a truthful self-appraisal.

How much damage this scepticism is doing is not difficult to see. One finds it in the disbelieving look on faces of people whose express policy seems to "reserve judgment" on everything they hear. The look seems to say: Don't trust what you hear and always read between the lines. This suspicious approach can become an official doctrine, and indeed is being put to use in commerce. The concern about "truth in advertising" suggests advertisers do not believe that telling the truth gets them anywhere and that the best way of assuring sales is to change people's perception as to what they really want. The appeal to reason is not expected to work; instead one attempts to mobilize the nonrational, subrational, or subliminal strata of the mind. The apparent "success" of the advertising industry is, unfortunately, taken as a confirmation that *all* interpersonal communication—private or public—is really manipulation or a mild form of brainwashing. Such manipulation, either practiced or experienced, is the enemy of the natural need to seek autonomy and harmony among one's beliefs, thoughts, desires, and actions. Autonomy is the condition of inner balance and sanity. Unless people foster the growth of this natural capacity for personal self-conception and the concomitant sense of being authentic authors of what they do, they fail to bring into existence values peculiar to my human form of life.

Autonomy as Self-Construction

To whatever extent persons are masters of their own destinies, they are in a position to build their personal worlds. It is quite appropriate here to invoke an esthetic analogy. Living a life is sometimes compared to creating a work of art, in which task the materials are external circumstances and personal talents, bents, and propensities. Out of those elements a person needs to construct something acceptable at least to himself. One of the reasons why esthetic analogy is appropriate is that treating a person's life as if it were a prefabricated structure is literally dehumanizing; humanity is reduced to levels discovered by me in preceding biological forms. I treasure humanity precisely because a person's life is a paradigmatic self-construction. As the artist is ultimately responsible for every brush stroke or every note, so every person can affect daily the shape his or her life takes. Like the sculptor or the painter, a person must decide at any given time whether to use a chisel or a hammer, oil or watercolor. Of course, like an artist's apprentice, an individual needs to go through preparatory stages in which the facts and the possibilities open to him are learned and explored. In some stages the process of self-shaping is more direct and intense. Even when a person finds that the work is more or less completed and believes that little can be done to improve it much, like a self-respecting artist he should not ignore occasions when life's canvas could be touched up a bit. No self-respecting person will rest indefinitely on laurels.

The search for autonomy is endless. Repeatedly it is prompted by a realization that some loose strands resulting from previous decisions could be arranged more neatly or allow for a new pattern to be woven. Fascinated with beckoning new possibilities, alert people are eager to have more time and more energy at their disposal to give these new options a try. The attractive thing about autonomy is that it need not be confined to one definite goal. Like life itself, forever sprouting new seeds, personal growth is open-ended. Remembering or keeping an eye on what they planted and what they harvested, they can see the pattern of their lives as capable of revision and improvement. New, unexpected perspectives may suddenly show. This openness to new possibilities is the reason why life is treasured up to the very last moment. My *conatus* is indefatigable, especially so when, in persons, it *knows* the value of what it is trying to bring into being.

❧ From Meaning to Mysticism ❧

The Birth of Meaning

In describing what happened to me when I discovered life I had to invoke the idea of meaning from the very start. Sentience and meaning were born together. Even a tiny organism senses its distinctness from its environment. To the extent that its feelings, its inner states, are affected by the surrounding, that surrounding is *meaningful*, in the minimal, rudimentary sense of the word. With the advent of language, when reference and self-reference became possible, the phenomena of meaning took on explicit, conceptual forms. Human beings, I have said, live in an *interpreted* world.

Conceptual awareness is a form of sentience, but a very sophisticated one. Before life there was no awareness, no consciousness. When life came upon the scene I began to wake up from my dumb, deaf, and blind state. What kind of world is there for an amoeba? Is it aware of *things?* Not likely. It *feels* its environment and in that sense is aware of its surroundings, but it does not *know* of what its environment is made up. To be aware of a thing one must be able to think it. But thinking involves the ability to refer, and of that only self-referring entities are capable. A thing, to *be* a thing with its separable reality, must be taken into awareness by a being that can separate, distinguish itself from that thing. From this it follows that things exist *as* things only for language-using, referring, and self-referring beings.

When one tries to speculate what the world is like for non-linguistic animals, one immediately stumbles upon the difficulty with the word "world." A world is an organized whole, at least an ordered collection of things. As such it depends on the ability to think, to form concepts. In that sense, when my biological existence was limited to pre-linguistic animals the world did not exist for me as a concept. I had no concept of a world because I had no concepts. My prehuman experience of meaning was dreamy, diffuse, and vague. My other forms of life are aware of their environment but it does not enter their consciousness as an *ordered* whole, located in space and taking place in time, having a history.

Since only language-using, referring and self-referring beings have concepts, only persons have concepts of things either existing as separable or united into collections. Uniting things into collections, seeing them *as* collections requires the ability to form universal concepts, that is, concepts that refer to a collection as a whole. It follows from this fact that only with the advent of persons I could acquire for the first time in my history a concept of myself *as a cosmos*! My meaning as a cosmos is mind-dependent and can be a focus of awareness only to the language-equipped minds of persons. This means that apart from persons I *have* no mind of my own. Since my awareness of myself as a cosmos is possible only through and in human beings, my infatuation with persons is understandable. Notice, however, that it would be a conceptual error to say that in my merely material, pre-life phase I was unintelligent or stupid. To attribute either stupidity or intelligence to an entity without a mind is senseless; that is, no sense can be given to such an attribution. Since I have no mind of my own (that is, a mind apart from or in addition to particular human minds), it is a mistake to attribute to me features that characterize only particular entities capable of consciousness.

Through the Eyes of Science

Through human beings, I can find out some interesting things about myself. They have discovered, for instance, that I am huge. That fact strikes them as significant — it puts them in their place. Compared to the size of the cosmos, not only a human being but also the entire planet he or she inhabits is a tiny speck in cosmic spaces. But consider how the *meaning* of this fact arises. I do not know how big I am. But persons do. Only intelligent beings who have invented the idea and the standards of measurement can be impressed by the bigness of anything.

Notice that both "matter" and "meter" are derived from the Sanskrit root *matr-*, "to measure." This bit of linguistic information suggests that the "material" world is the world as measured or measurable. This means that only for a mind can spatial and temporal dimensions be mind-boggling. As I have already mentioned, the knowledge of physical dimensions is a function of self-consciousness, hence it is self-consciousness that enables human beings to notice the *contrast* in size between what they are and what the universe is in that respect. It is they who are impressed by the immense sweep of galactic space and cosmic time, in which the entire human history appears insignificant.

Notice, however, that the perception of the contrast, the marvel about the distances between galaxies and the immense physical forces generated inside my stars and suns, is something *added* to these distances and forces. Persons are the eyes and ears and minds through which I learn facts about myself. Prior to their discovering and formulating the laws which govern the configurations, distribution, and behavior of physical energy in the cosmos I was blind, deaf, and dumb. People should ponder the words of a neglected but profound and poetic thinker, Kenneth L. Patton: "The world has forms, colors, light and dark, intricacy, movement, vastness, but it is not awesome; it is we who are awed."[5]

The fact is that human consciousness has produced the first comment ever made about me as cosmos. But that comment need not limit itself to the perception and measurement of distances, spatial and temporal, and to sophisticated descriptions of the ways in which my matter behaves. The standards of scientific theories are but some of the standards human beings can employ in determining the meaning of anything. Scientific theories and measurements are essentially *quantitative* in nature, and to the extent that they yield information and knowledge they are of value to human beings, of great value no doubt. Factual knowledge, however, is only one of the values which human beings cherish, and they have other values as well, values that determine the *quality* of human life.

Qualitative Standards

Measured by standards regarding quality, my immense quantitative dimensions may suddenly pale into insignificance. One can cogently ask whether it is better to last for a long time in a state of nonconsciousness or to exist for a limited time in a state of some awareness and knowledge. The mere asking of this question breaks the spell of space and time. Having discovered the human form of life, I have realized that there is no reason to be impressed by the tyranny of quantity. My quantitative cosmic meaning is swallowed up by that qualitative value dimension discovered by persons. Many of them begin to understand that to point to the sheer size or to the long-lastingness of anything is not to point to any special distinction. There is no reason to worship anything just because of its dimensions. A misguided deference to big things and long-lasting processes ignores qualitative distinctions. It makes sense to regret the passing away only of *deserving* things and processes, not of those that lack significance and value. I grant fully that there is no reason to regret or to rejoice in the destruction or the passing away of one of my galaxies and the replacement of it by another, if neither the one nor the other displays anything *worth* preserving, that is, if the qualitative standards cannot be employed to differentiate between them.

Repeatedly, in their intellectual and spiritual explorations, human beings return to the question: what is the universe in itself, apart from us? Does it contain forces, tendencies, or purposes of its own? What can we say about the ultimate nature of the cosmos? If my reminders assembled so far are cogent, human beings have reasons to resist the temptation to think that I am meaningful in some profound but nonhuman or superhuman sense simply by virtue of my sheer physical magnitude, my astronomical expanse in space and time. By now they should get over the shock of the scientific revolution which disclosed to them the vast, incomprehensible age and size of the cosmos. Only when they are awed by my spatial and temporal dimensions can they be depressed by the thought that the earth is but a tiny speck in the cosmic scheme, displaying the temporal drama of life and culminating in an even smaller stretch of events of human history.

People can cease feeling overwhelmed and insignificant when they realize that it is *persons* who introduced me to the value of knowledge. I am quite serious when I suggest that human consciousness is a way in which I find out something about myself. The very idea of *finding out* comes into being only with persons. This is one reason why persons bring something uniquely valuable into my cosmic spaces. The vast expanse of matter that in most (perhaps all!) galactic contexts is able to produce an awesome display of physical energy is literally mindless. What is going on inside the sun, or on the surfaces of other planets, is certainly complex and powerful, but nowhere does it display meaning. To quote Kenneth Patton again:

> The sun has no reveries, sentiments, nostalgia.
> We imagine the sun-spot as a giant orgasm, a
>   rank human fantasy, an obscenely human distortion.
> The sun-spot is a chemical explosion, coarse,
>   unreflective,
> It does not relate to us, except to disturb the weather.
> The miniscule synapse of the neurons of a brain is
>   more significant than the tirade of stars.[6]

Meaning is found in scientific know-that and technological know-how, impressive achievements indeed. Nothing in my cosmic expanse *knows* how old I am, by what laws, micro and macro, I am governed, except persons. Science is a brand-new phenomenon in the cosmos. The satisfaction of knowing the nature of things is ranked among the highest human goods. When Aristotle observed that all men by nature desire to know and that there is no higher contentment than that derived from contemplation, he struck a splendid chord, unheard of in my entire history up to that time.

Science, however, is but one of the human mind's possible uses. Science depends on creativity and imagination, and both endowments are displayed in other areas that make human lives meaningful. Besides the satisfactions derived from *knowing* things, persons also derive satisfactions from *changing* things in the light of their understanding of what is worth bringing into existence. Besides the true, there is also the good, and the beautiful—and all three afford them the opportunity to be rational.

What is introduced into my cosmic spaces by intelligence enlisted in the task of devising admirable ways of living is no less important than the use of intelligence in natural science. Among my mankind's heroes are not only scientific giants, but also those who have contributed toward devising satisfactory and satisfying ways of organizing human life—on its economic, political, and broadly cultural planes. People rightly honor those who have given them the rule of law, from Hammurabi to Plato, from Jefferson to Lincoln, and those who were willing to give their all to promote humanly desirable ideals: Socrates, Joan of Arc, Giordano Bruno, Admiral Nelson, Patrick Henry. If one thinks of human lives not in quantitative but in qualitative terms, some of them deserve to be regarded as significant and unprecedented cosmic events. From the point of view of meaning and value, momentous events in human history are also landmarks in my cosmic history.

It would be a mistake, however, to think that only historically weighty events and especially creative lives affect the status of human values on earth. Exceptional lives merely underscore the dynamic function of standards to which all human beings can respond. As long as they live in the light of standards, using them to inform, in the sense of giving form to their actions, they introduce special values into the cosmos. The point I am stating is that the way the world is experienced makes a difference to it; it becomes a different world.

Take, for instance, the phenomenon of spring. What *is* spring, you might ask. What belongs to its full characterization? Not just the physical and biological phenomena, such as earth warming, sap rising, flowers blooming, birds singing. Part of spring in the *celebration* of spring. To come into itself, to reach its humanly extended potential, spring requires the feelings appropriate to it. The spring, one might say, *needs* human beings if it is to be what it potentially can be. Without them, it does not quite come into its own, does not realize its full possibilities. Likewise with the stars people see at night. By espying them and by perceiving their order, both in astronomical calculation and in poetic response, they add something to the character of nature, namely by knowing and by esthetically appreciating that character. The lady who wondered how the astronomers ever discovered the names of stars may be simple-minded, but behind that simple-mindedness there is a germ of philosophical wonder. The physical world without physics is a diminished entity. An undiscovered and unformulated law is still a law, but somehow it seems deprived when it is not known. In that sense, it does not come into its own; its formal identity does not exist. Being discovered and formulated by a physicist is, so to speak, its good fortune.

The Fruits of Understanding

What distinguishes human beings from other animals is the ability to understand, to express themselves in language and in art, to appreciate, to mourn, to celebrate, to lament, to enjoy, to praise. Things themselves are dumb, but their meaning can be captured in human experiences. In his poem *Duino Elegies,* R.M. Rilke cites a string of capitalized words: House, Bridge, Fountain, Gate, Jug, Olive Tree, Window, Pillar, Tower. These carefully-chosen words, in order to be understood in their full meaning, require the human context. A multitude of meanings must be read into every one of them to get at their full significations. It would take a historian, a scientist, a poet, a philosopher, and perhaps an anthropologist to show concretely and informatively how these sorts of things are intertwined with the story of human development and culture, of aspiration and achievement, both on the large stage of life and in individual personal careers. There is no denying that very special *additions* have come into the world with the entry of the human mind into it. Think of Plato's creation myth, according to which the wholly unintelligent and unintelligible chaotic matter *becomes* a world only with the infusion into it of intelligible, mind-governed Forms. When persons permeate their lives and surroundings with thought and feeling, with intelligent and creative designs, they rescue me from anonymity and meaninglessness. But I do not find that the ideas in terms of which I am experienced are intruders; on the contrary, they enable me to exist objectively at the level of meaning.

Immanuel Kant claimed that the world and the self, as meaningful structures, arise together. This mutual interdependence allows a widest possible interpretation. Just as the application of theoretical concepts to human experience can yield objective knowledge, so the use of standards of action shapes people's cultural and esthetic pursuits. What is called the life of the mind has its home in concrete natural setting and gives expression to varied needs and interests of the human race. The emergent form of life displaying itself on the surface of one of my relatively small planets manifests qualities that have no counterpart anywhere else in my immense cosmic spaces. Should I discover a form of life superior to the human, such a discovery would not necessarily reduce human achievements to nothingness. They would still retain their intrinsic value, which would have to be acknowledged side by side with whatever other valuable actualities should come into my being. A discovery of another impressive form of life would only show that there is more in me than is dreamed in my human philosophies. But creatures arriving on earth from other planets would be narrow-minded and probably unwise if they chose to ignore or to spurn those philosophies. They might learn something from reading Plato, Locke, and Kant, Sophocles, Shakespeare, and Goethe, provided they *could* read.

Stories which the earthlings could tell to creatures from outer space would convey incredible thrills. What must it have been like for Einstein to come upon the tracks of his revolutionary theory of relativity? For Beethoven to compose his Ninth Symphony? For Dostoyevsky to agonize over the next chapter of *The Brothers Karamazov*? For Pericles to plan the building of the Acropolis? For Columbus to sail doggedly toward the New World? For Lincoln to face the Gettysburg audience — living and dead? For Pasteur, Curie, or Salk to make their discoveries or for Edison and the Wright brothers to realize their inventions? I still glow with pride when a person recalls Neil Armstrong's words when he landed on the moon. To comprehend the grandeur of that event the guests from other worlds would have to rediscover all of science laboriously acquired by the human race. Only after having done so could they share Carl Sagan's excitement that accompanied his narration of the "Cosmos" TV series.

Piety

There is an interesting ambiguity in the concept of meaning: sometimes it is used as opposite of senselessness, having no sense, but it may also be taken as equivalent of "meaningful," as in the expression "This means something to me." "Meaningless" may mean "having no meaning at all" or alternatively "having no significance or value" — as in the locution "This event had no significance." These two senses of "meaning," taken as opposite poles of a spectrum, are ideal abstractions. But all forms of sentience, including thinking , have meaning in the sense that they signalize *involvement* in the world, a fundamental type of *caring,* being oriented toward the object of thinking. Philosophers have a special term for this feature of thinking: intentionality. They say, metaphorically, that every thought *intends* its object. Perception involves intentionality. In an act of perception the mind and the perceived object are not in an external contact but in fact form a union, a mutually-dependent, mind-ordered whole. The blueness of the sky is not separable from a perception. When one looks at the sky on a cloudless sunny day one is, of course, not looking at an inner sensation existing only in the mind, but one sees the sky itself as it appears to the perceiver. To perceive a thing or event is to bring it into a relation it itself does not suspect. It is that *relation* that is meaningful, not the things themselves that appear in that relation.

Since it is the thinking person that brings the relation into being, meaning can be said to be mind-dependent. But in exhibiting intentionality a person pays attention to objective surroundings, and "objectivity" is a respect for the independent nature of the objects of thought. Even the phenomena of memory and imagination, in which "the object of thought" is a person's subjective content of mind, presuppose a previous contact with or a possibility of real objects of experience. A memory report is unsuccessful unless it is traceable to a real event, and imagination is defined as a *contrast* with what is factually the case. Thus the primary sense of meaning is *objective* meaning, a situation in which, as Kant claimed, the mind and the world arise *together*.

A consequence of this fact is something that could be called "the ethics of belief." The phrase was used by W.K. Clifford to argue that it is wrong to believe anything on insufficient evidence. Clifford's point is a special case of a more general requirement which bridges the supposed gap between "truth" and "truthfulness." Truthfulness is a virtue of a person who respects truth. To respect truth is to be guided by what is in fact the case; to respect facts, to let things as they are be the guide to what one believes. The delight in knowing things in this way is eloquently expressed in Democritus's statement that he would rather discover one fact than be the king of the Persians. In saying this Democritus brought out the special nature of the value introduced into me through human consciousness. The world comprehended is a special kind of world, and a mind can rejoice in the fact that the world *is* intelligible. Such a mind will respect the conditions of intelligibility, which include a respect for facts. This respect will also include a virtue which Socrates called *piety*.

In Plato's dialogue *Euthyphro*, Socrates is trying to elicit from his self-confident interlocutor a definition of piety. Euthyphro does not get very far under Socrates's merciless questioning. In the end Socrates suggests that piety may be a kind of justice. Justice enjoins to give each being its due. If that being is God, then that being ought to be granted what is due to God. If that being is a human person, that being ought to be granted what is due to him. Socrates does not specify *what* is due in each case, and the dialogue leaves this question unresolved. Nevertheless, the phrase "is due to someone" is felicitous. Its normative thrust becomes clearer when it is put negatively. One can ask what happens, for example, when a person is not given what is due to him.

Plato develops the idea of piety in a moral context and considers it as a moral concept. To treat the interests of others as *deserving* recognition, as being objectively valid, is a form of special piety. Its emotional concomitants are empathy and sympathy. To feel *with* another, to pay attention to what he or she is achieving or creating, is an expression of piety. To applaud and to cherish the excellences and accomplishments of others, to give them hearing, support, recognition, and encouragement, is to celebrate phenomena that are objectively good and worth celebrating. To turn one's back on them is not to give them their due. Not to applaud a performer after a splendid concert is a form of injustice; even to be a boor or a prig is to be deficient in piety, for such dispositions ignore values that deserve appreciation.

Piety, then, as a disposition that gives all deserving phenomena their due, has a wide scope of application. It is present in the creation and appreciation of science and art, in the sedulous or inspired exploration of meanings discovered, amplified, extended, echoed, and re-echoed, in myths and theories about the universe, life, and human history. To read a great poem with understanding is to reach deeply into the depositories of language as it structures and illuminates the experience of humanity, observed and commented upon by its talented representatives. A sympathetic journey into the recesses of the human mind, heart, and soul is an act of true piety on the part of those who create and those who cherish human ventures into history, philosophy, science, poetry, religion, and art.

Piety need not be restricted to the human context. The disposition to let things be, to give them their due, may cover a wide territory. It should not be forgotten that human beings share something very fundamental with all other living beings, namely, sentience, an intrinsically good phenomenon. But if sentience is good in itself, then piety dictates that it should be respected everywhere. Whether the respect for animal life is explicitly moral may be questioned, especially if by morality is meant a reciprocal relationship, recognizing each other's rights and obligations. Not being able to use language, animals do not have the concept of obligation, and hence, one cannot expect them to understand what it is.

This observation does not mean, however, that it is right for human beings to ignore the fact that animals can experience pain, deprivation, suffering. Taking an interest in their well-being is, therefore, also a kind of piety. Albert Schweitzer's name for it was "reverence for life." This form of piety fits Socrates's definition of it as giving an entity its due, not depriving it of what in justice belongs to it. Life belongs to living creatures, and it is impious to take it away from them. Even killing animals for food turns out to be a questionable activity when it is examined from this perspective. The fact that for centuries man has arrogated to himself the right, often sanctioned by religion, to dispose of animal life at will for his own benefit appears reprehensible to many critics who condemn vivisection and laboratory experimentation which, like wholesale factory farming, involve extremely cruel treatment of animals. Some philosophers advocate "animal liberation" and vegetarianism. The argument that all living things have their own lives to live and deserve respect is extended even to plants, and some philosophers are prepared to claim that even trees "have standing," that is, should be accorded certain rights.

Even though these views are not universally shared, they nevertheless gain more and more adherents, which does credit to human sensitivity. The varieties of piety show that it is indeed an open-ended concept. Just what are its reasonable limits is not easy to determine. What about mosquitos? — a person might ask. Is it all right to kill them? The Jains of India are true to their religion when they wear masks so as not to swallow small insects inadvertently, but still they forget about bacteria destroyed by their bodily processes. Similarly, Schweitzer's "reverence for life" was surely at odds with his medical calling. I do not fault human beings for their hostility toward harmful bacteria and some other forms of life such as ants, wasps, and roaches. They are rightly sceptical of whether in destroying such forms of life they are inflicting much pain or suffering. Out of respect for their own lives they ought to defend themselves against dangerous or threatening non-human intruders, tiny or large. With these legitimate caveats, I nevertheless hope that people will abide by the principle that causing unnecessary suffering is wrong and that they should not interfere with any life that is doing no harm.

Mystical Feelings

The widest form of piety comes to expression in *mystical feelings*. They arise when caring and letting things be merge into a sense of all-inclusive meaningfulness. In mystical feeling there is a unification, a mental telescoping of my entire cosmic sweep. Like other forms of piety, it is not just a subjective feeling but arises from objective states in the world. When persons view the total scheme of things valuationally, and try to give all things their due, they are embarked upon mystical experiences. They acknowledge the boundless fertility and creativity of life in my cosmic vastness. They cherish the spectacular achievements of the human race and take an interest and even delight in the interests of others. In other words, they try to admire all things in their actual and potential beauty. They feel intensely that I am not a collection of value-neutral facts.

Mystical feeling arises from realizing that an attentive survey of what there is makes *all* of my possibilities a legitimate object of a person's piety. To give things their due, nothing must be slighted or ignored. One instance of beauty, Socrates claimed in the *Symposium,* recalls or calls attention to another, and where can you stop? To be just and fair to all of them, one must try to accommodate them all. This calls for expanding one's time-consciousness and for seeing things, as Spinoza put it, under the aspect of eternity.

Although mysticism is usually discussed in deeply philosophical or religious contexts, it need not be confined to them. Mystical experiences can be had in ordinary circumstances, when the flow of time is condensed into a pure now. An ordinary sunset or an average seascape may bring on a minimystical experience, when the passage of time seems irrelevant and the momentary perception is telescoped into the vision of a comprehensive unity of all things. In some experiences the extraordinary sense of unity need not be dominant. Instead, the person who has them may find other metaphorical expressions more appropriate and will speak of depth, mystery, or ecstasy without being tempted to embrace supernaturalism. All of them open new "doors of perception" and put a person in touch with unsuspected possibilities. In all of them the basic principle of piety is at work: things are given their due, acknowledged in their reality as admirable in themselves, and hence eligible for inclusion in a holistic vision. To be moved by such a vision is to show forth ultimate piety.

Religious mystics report on the absolute enlargement of the self, which they attribute to merging with God, with all reality. The interdependence of the two merging entities is expressed by the German mystic Meister Eckhart as follows:

> The eye with which God sees me, is the eye
> with which I see Him, my eye and His eye are
> one. In the meting out of justice I am
> weighed in God and He in me. If God were not,
> I should not be, and if I were not, He too
> would not be.[7]

Simlarly, Nietzsche's Zarathustra greeted the sun with these words:

> You great star, what would your happiness be
> had you not those for whom to shine? For ten
> years you have climbed my cave: you would have
> tired of your light and of your journey had it
> not been for me and my eagle and my serpent.[8]

From such encounters, there arises the sense of the totality of things as absolutely admirable. A unified whole comprising in me all good things and events is momentarily grasped and celebrated. This act of ultimate piety, giving every thing its due, attaches the self to its contemplated object and sees both, itself and the object, as constituting an indivisible and intrinsically admirable unity. This act returns the self to me and celebrates both itself and me at the same time. The circle is completed.

The completion of that circle signalizes the intimate connection of personhood to my cosmic reality. Thus, late in my cosmic calendar, I may find a unifying expression in some experiences of persons. It is no surprise, therefore, that upon perceiving this unity with me they personalize me. In doing so they are affirming the special value of personhood while acknowledging their physical rootedness in my star-stuff and their belonging to the cosmic totality that makes them possible. Even if very recent, personhood is to date the final product in my discovery of life. Although that product is creatively open-ended and oriented toward the future, persons are justified in feeling a kinship with my whole cosmic process. The sense of this connectedness can even palliate the sting of death. As the most inclusive form of piety, mysticism enables persons to feel timelessly at home in my cosmic spaces. They rightly feel that they play a special, privileged role in my indefatigable quest for meaning and value.

❧ From Happiness to Blessedness ☙

Personhood and Self-Conception

Before turning to another important transition in the growth of life toward full personhood, let me review the key elements of that growth. At first slowly and laboriously but gradually gathering momentum, I have moved from the star-stuff stage to living organisms and then to human beings. The distinguishing feature of humanity is personhood, the ability to perform acts justifiable by appeal to objective standards. Persons are deliberate creators and sustainers of values, in contrast to other animals who, lacking rationality, cannot choose values in accordance with such standards. Persons can make choices for the sake of the good and justify actions in terms of values they bring about. One reason for regarding persons as the most valuable form of life is that their very mode of life is value-creating. I am almost tempted to read the meaning of "valu*able*" as *"*able to create values." This ability is most prominently displayed in the quest of personal integrity and autonomy. Persons strive to structure and to harmonize various candidates for values, thus producing admirable individual lives and impressive social structures—cultures and civilizations—through the efforts of creative individuals. For this reason autonomous activity of persons can be thought of as a value of values or a second-order value, since it is the condition of all first-order values realized in daily life. Among such values are morality, piety, and mystical feelings.

Persons who value their experiences, that is, find them good, welcome, satisfying, admirable, are said to be happy. Happiness is a state of mind which a person affirms as worth having for its own sake. When people say that the *pursuit* of happiness is the goal of life they acknowledge the dependence of happiness on what a person *does*. And what a person does in life depends on that person's self-conception. That there is a connection between happiness and self-conception was clearly shown by David L. Norton in his book entitled *Personal Destinies.*[9] Norton shows that the contemporary meaning of "happiness," the core of which is the feeling of pleasure (no matter whether determined quantitatively *á là* Bentham or qualitatively *á là* J.S. Mill) does not capture the full meaning of the word "eudaimonia" from which it was derived. The Greek meaning of that word is far removed from what it came to mean in its English translation: "happiness." Norton points out that "eudaimonia" should be seen as a person's state of mind derived from following his own "daimon"—the particular genius or spirit that that person gradually discovers as his or her proper self—and then acting in a way that is consonant with that particular genius or spirit. "Eudaimonia" is in part a function of taking into account one's particular traits, bents, talents, capabilities, strengths, and weaknesses. Given a certain self-conception, some things that a person does or tries to do simply won't fit, they won't give *this* person happiness, even though they would give happiness to some other human. There must be a certain match between one's self-image and the pleasures that come one's way. The awareness of a mismatch often comes only in retrospect. Most people recall having responded to pleasures which they later see as childish, adolescent, or sophomoric. Perhaps these pleasures were appropriate for the stages of life in which they occurred, but the growing maturity does not quite see them as manifesting the best judgment, reached only with growing experience.

Undoubtedly, with the growth of experience and knowledge, including self-knowledge, the self-conception of a person may shift. Typically, it does not shift haphazardly, just in *any* direction. Throughout their development and maturation people acquire a certain sense of personal identity, and as the result of accepting and cherishing it, they judge the value of potential satisfactions in terms of what kind of persons they believe themselves to be. Being the person that one is, one finds some things attractive, satisfying, fascinating, or, conversely, boring, distasteful, indifferent. The choice of things people seek as good in themselves clearly reflects their bents, dispositions, and talents. It is also conditioned by their skills, abilities, and proficiencies. Not surprisingly, people seek satisfaction in work, in performing tasks of their vocation or profession. The satisfaction increases when they find these tasks challenging, absorbing, exhilarating. They are fortunate when their daily work allows them to be creative.

In the first sentence of *Anna Karenina,* Tolstoy observed that every happy family is happy in the same way, but that the unhappiness of each family is unique. This famous comment may be true with regard to families, but it is at most half-true when applied to individual persons. For if it is the case that a person's happiness is conditioned by what that person is, then no two seemingly identical satisfactions are alike. There is a sense in which each person *creates* his or her own happiness. To say this, however, is not to attribute to human beings a heroic capacity to conquer every pain and to transform every adversity into an opportunity. There are situations in which misery and unhappiness are as predictable and understandable as hunger or thirst. But the range of human reactions to what offers itself as a possible satisfaction is infinitely varied.

Happiness and Education

Persons' reactions to what experience brings them is a function of education. Consider an example from the realm of music. To an untrained, musically-uneducated ear, classical music may mean little, if anything. It takes at least some exposure and sympathetic guidance before one can derive satisfaction from a musical experience. There seems to be no upper limit as to what that experience can bring. A person with a thorough musical education will perceive in a performance features that escape the greater part of the audience. Take any audience at any symphony concert. What is going on in the minds and emotions of each listener is a function of so many variables that the question whether each hears "the same" piece of music is not pointless. The "input" is the same, in the sense that the same score is played (even though here the different "reading" and interpretation by the conductor and the musicians justifies saying that each performance is unique), but what happens on the receiving end varies from person to person. Some differences here are due to musical training and previous exposure to music; others to native musical talent or absence thereof; and still others due to inattention, fatigue, distraction by physical or emotional factors. One cannot enjoy Beethoven while suffering from a throbbing toothache or thinking hard where to take one's date after the concert.

What happens in a concert hall analogously applies to any experience of happiness. As the personal musical experience varies from person to person and from time to time, the experience of happiness is never the same. What actually happens in any case is in part due to objective factors which at the moment cannot be controlled and in part to the subjective contribution of each person. Just as at a concert one can admonish oneself: stop daydreaming and pay attention to what is coming your way, so in various life situations one can tell oneself: allow each experience a chance to give you what it has to offer in terms of positive value. It has been said of Arturo Toscanini that he peeled an orange with the same concentration with which he conducted a symphony. Many good things in life do not get realized because of inattention and distraction; a person's mind is not where it should be, namely, where that person is.

There is such a thing as a capacity to be happy, cheerful, receptive and sensitive to joy-creating situations. Some people seem to be born with a happy disposition, innate cheerfulness, or a sense of humor. One rightly considers possessors of such natural endowments as fortunate, and I sometimes wish that I could distribute these endowments more universally and more evenly. The same goes for natural beauty or good looks, although in this regard people tend to believe, with some justification, that they can improve on nature. This belief accounts for a thriving cosmetics industry and beauty salons. People also begin to realize that shyness, moroseness, and depression are not altogether beyond their control. They can distinguish, for instance, between temporary and chronic depression, and devise techniques to overcome shyness. Some illnesses are psychosomatic or self-induced, and laughter, they say, is the best medicine. Norman Cousins ascribed his recovery from a serious physical illness at least in part to deliberately-induced laughter.

Happiness and Temporality

Autonomous, healthy, and effective individuals tend to radiate happiness. They know that, like autonomy, happiness is intrinsically connected with human temporality. They realize that every single moment is a focus of three-dimensional temporal span. No matter how intense and absorbing is the pleasure of a given moment, they see it as belonging to a stretch of other moments. Happy people seem to see a connection between every moment of their lives and the whole of their lives. They believe that the full appreciation of the meaning of an intensely happy moment tells them something about their lives as a whole. While experiencing such a moment people can say to themselves: *I* am the person to whom this happy moment belongs; in experiencing it, *I* have been fulfilled in some important way.

When people ask themselves whether they are happy they reproduce in memory events and incidents which they treasure because these express what they find truly valuable in their whole history. They are likely to think of golden moments, achievements, ecstasies and good times. These "highs" need not be hedonistic in character. They may include moments of satisfaction derived from performing acts of kindness to others, of unselfishness, of conquering unworthy impulses. People then will remember how they used talents in effective and rewarding ways, got rid of bad habits, managed to perform painful tasks, or showed self-restraint in provocative situations.

Correspondingly, in looking forward to the future, including old age, people think hopefully of moments and stretches of time which will make that future good, even though they realistically expect some rain to fall into their lives as well. Emotion and thought are engaged both retrospectively and prospectively, aiming at their fullest harmonization. To experience a pleasure that appears disconnected from self-conception is disconcerting. It seems not to fit and is viewed as a disruptive intrusion, preventing happiness from occurring. In a sense, for happiness to happen, the persons experiencing it have to happen as well.

Blessedness: Affirmation of Value

Once more I would like to return to my original discovery of life in order to call attention to what is special about it. Life as such is the affirmation of value. The earliest mode of this affirmation is *conatus,* a feeling of acceptance by an organism of welcome states, that is, states that maintain its vital integrity and well-being. The affirmation of this good is at first diffuse and inarticulate, gradually transforming itself into instinctive mechanism of survival. When life reaches the human stage, the affirmation blossoms into a conscious pursuit of particular forms of goodness articulated in standards deliberately adopted and transmitted through learning from generation to generation of persons. When people find that something matters, is worth caring about, they proclaim it to be good.

Interestingly enough, the Latin verb *benedicere* literally means "to proclaim good." Most frequently used in a religious context, it refers to an act by which something good is bestowed upon a person. To benedict is to bless, to bring a person into the presence of most desired goods—well-being, prosperity, happiness. Indeed, "eudaimonia" is often rendered as "blessedness," perfect contentment of heart and mind (Funk and Wagnall's). The Slavic version of "benediction" or "blessing" also means literally "to declare good" *(blogo-slavit)*.

Blessedness is a comprehensive concept, embracing all positive features by means of which various thinkers, philosophical and religious, have tried to characterize the fulfillment of life. To be reassured that one's personal existence matters in an absolute way is to be in a state of blessedness. To be in such a state is to be suffused by the conviction that life is ultimately good. That conviction is a result of a certain kind of knowledge or wisdom, in the light of which a life is lived. As with other kinds of knowledge, blessedness is not something of which a person must be constantly conscious. Rather, it can constitute a steady background of belief which can come to the forefront of one's attention when the occasion calls for it. Once acquired, however, the conviction that one's life matters in this inalienable way can surface in one's consciousness in many different contexts.

One such context may be an occasional reflection on the fact of one's own existence, of having been born. If, as I have claimed, persons are beings through which I can see myself as a meaningful whole, then every upsurge of human consciousness through the birth of a person is both a triumph and a privilege. It is a triumph for me and a privilege for the emerging individual. Every human birth is a joyful event and calls for a celebration. No less than life as such, I see every birth as a miracle, as a victory of life over non-life, over the barren silence of the merely material, literally meaningless world to which I ignorantly confined myself for such a long time. Persons' lives allow me to speak out what I can be, in an endless, inexhaustible sequence of unique personal careers of individual human beings.

The state of blessedness experienceable by persons is far from being one-dimensional. "Blessedness" is an umbrella term that stands for a variegated joyful wisdom, and it is a mistake to reduce it to a formula. As the most refined, complexly structured organisms, human beings are far from simple. Why should their happiness be of only one kind? Shakespeare knew about this infinite variability of humanity when he told and showed his audiences: "I'll teach you differences."

The Pursuit of Blessedness

In an individual life, blessedness, like wisdom, seldom settles into a steady possession; it can move across the entire horizon of a person's experience, intermittently illuminating its course. It can well up in peak experiences, in sudden exhilaration, in flashes of insight, in encounters with truth and beauty, in stretches of contentment, in mystical feeling. For a being that seeks its fulfillment in the framework of time, hour by hour, day by day, year by year, blessedness cannot be continuous. But since it is a state of mind, it requires a proper attention, a *presence* of mind of the one who wishes to experience it. Alas, all too often people lapse into drowsy inattention. The dividend for avoiding such passive somnolence has been eloquently described by Stephen Lackner, whose advice for humanity I fully endorse.

> If we could only use to the utmost those hours of wakefulness, looking out with wide-open eyes, plumbing as clearly as possible our own inner depths, making each brain cell and each muscle function joyfully and actively; if we could only perceive the whirling crowds around us, together with ourselves, as parts of an integrated organism called humanity; if we could seek and enjoy nature as help and wonder; if we could acknowledge the full value of the outer world, which depends on the perception of our senses and whose quality must disappear again when we close our eyes and ears for the second sleep; if we could maintain a state of heightened wakefulness at least for the decisive periods of our own time: Then our whole life would be worth the trouble—all the trouble.[10]

Persons who find life to be essentially good realize that the state of their blessedness is in part their own doing. They are right in regarding personal values as having cosmic significance. Since my meaning is in human interpretation, human blessedness is a crowning result of that interpretation. All human dignity is in human thought, said Pascal. His declaration can be expanded by adding that thoughtful personal subjective responses to my objective conditions and possibilities transform me into something potentially glorious. I reverberate with that glory when I am cherished and celebrated as a carrier of values. Human beings articulate and disseminate these values in their untiring pursuit of what is reasonable, sound, noble, and beautiful.

That pursuit is not always crowned by success, and blessedness is often blocked by adverse conditions. I do not deny that human life, like all life, is subject to accidents and blind natural catastrophes. Moreover, in contrast to less developed forms of life, persons are capable of deliberate, gratuitous evil, frustrating all admirable designs. It is also no secret that the content of a person's life depends in part on good fortune. Some rain falls into every life, and tragedy is not easy to avoid. Nevertheless, barring conditions beyond their control, people can determine the frame of mind in which they develop their capacities and face their circumstances whatever they are. The advent of blessedness, intermittent as it is, can be prepared for by patience, effort, and perseverance, and above all by thinking appropriate thoughts. The cynics, sceptics, and scoffers are wrong when they deny that one's outlook on life is of one's own choosing. The possibility of such choice is opened up by the realization that the meaning of events is not dictated by external forces but depends on human interpretation, which often allows an actualization of blessedness.

The Drama of Personal Destinies

In saying that life in personal form is my favorite mode of being, I am taking note of the obvious fact that the most interesting drama of life is played out at the level of human action and history. Recall that the meaning of the Greek word "drama" *is* action. Every day and every moment the history of humankind takes on its actual shape through decisions of persons. Not without reason did existentialism catch popular attention by declaring that human beings are entities that put themselves in question. But this verbal formula invites a natural sequel, since a question normally calls for an answer. Because what is being questioned in this case is the whole entity, the individual person, it is quite proper to add that the way that person *lives* constitutes the answer to that questioning. Since the question is large, so must be the answer. Both the asking and the answering can take a lifetime. In the light of the account of personhood as I am presenting it the existentialist slogan can be expanded: human beings are entities that put themselves in question *and* are called upon to provide an *answer*.

Existentialism is a philosophy that emphasizes the sense of uncertainty as to whether human life has any direction, goal, destiny, or ultimate purpose. An analogous uncertainty can be discerned in contemporary science. Scientists increasingly are beginning to wonder whether their theories describe or refer to any independently existing reality. The outcome of scientific investigations is "infected" by concepts, hypotheses, and measurements hatched in the scientists' minds. The output depends on input. The world that emerges for human inspection is a world partly invented by people and "doctored" by anthropomorphic theories. In addition, when the changes in scientific claims are examined, they do not disclose a steady, consistent, logical path. As Thomas Kuhn has shown in his book, *The Structures of Scientific Revolutions,* scientific paradigms outlive their usefulness and are replaced by entirely different ones, thus introducing revolutions into "normal science" and instigating brand-new types of investigation.

Such claims naturally raise the question whether there is *progress* in science. Kuhn's answer is that if one conceives of progress as movement toward a pre-existent goal, then one cannot say that there is progress in science. But if one instead asks whether something has been learned about the nature of physical reality since the beginnings of the scientific enterprise, the answer is yes. If progress is measured in terms of where the contemporary state of science comes *from*, one can point to successes. If, on the other hand, one thinks of progress as movement *toward* some goal, one doesn't know whether scientists are progressing because they have no independent way of determining what that goal may be. They discover what is possible to know only as they go along.

Kuhn's suggestion seems sensible enough. It does not imply irrationalism or relativism because one can point to real successes in solving particular problems, removing specific uncertainties, and correctly describing phenomena that call for an explanation. Many a theoretician has made especially significant contributions to scientific inquiry—some of them have received the Nobel Prize for doing just that. Many others serve as links in a long chain of hypothesis formation, experimentation, and testing. Even Newton claimed that he stood on the shoulders of giants. Each discovery, invention, and advance must be evaluated from within the contexts to whose clarification it has contributed and in terms of what specific values, theoretical and practical, it has helped to realize. Not to be ignored is the fact that the practical applications of science secure values sought by agriculture, industry, medicine, and technology.

The open-endedness of science is paralleled by the open-endedness of personal human destinies. The existentialist thesis is on the right track because it is natural for human existence to be an issue to itself. I am conscious of the fact that the development of humanity is an unfinished business. I must also confess that I have no blueprint according to which the development of humanity must proceed. Instead, the work of determining what humanity is to be is in the hands of particular persons who put themselves in question and who provide an answer in terms of the lives they choose to lead. In this way each person participates in the formation of humanity, and in that sense is responsible for the course the career of humanity takes. How that career is understood and interpreted shows up in what possibilities are envisaged by particular persons and how they are enacted in societies of which they are members.

There is an analogy between the course of a human life and a development of science. Science defines its domain, chooses its problems, and devises solutions for them. Similarly, many problems of life are set by history, by former discoveries, failures, and anomalies. Much of the time the problems one encounters are familiar or are variations on a common theme. In that sense there is activity arising from "normal common nature" as there is work in "normal science." But many of the tasks and situations are either entirely or relatively new; and this is where persons must do fresh thinking and where their skills and talents are put to a test, at times requiring a shift in approach or paradigms.

An illustration may be helpful at this point. Consider what happens when you face the task of rewriting a paper. Your aim is to improve on the first rough draft or on the original version. How good the rewritten version turns out to be depends on its distance in quality from the version with which you began. The quality of the finished product depends not only on the content of the original version but also on what is brought into the revision by the effort to improve it—in terms of fresh perceptions, new ideas, clearer expression, more elegant style. These features are *added* by the writer in the process of performing the task, and the final judgment turns primarily on the actual contribution made by the writer to what he started with. This illustration sheds light both on changes in science and the changes taking place in a person's life. In both cases, the value of the outcome is determinable by examining the particular context and the background against which the task—solving a problem or living a life—is performed. In the case of scientific investigation a bit of scientific reality is revealed; in the case of living a life, a bit of humanity is unfolded. If what is unfolded in a life is admirable in itself, that life deserves to be blessed and celebrated. In achieving meaningfulness it bestows meaning on me.

3

*Barriers
to Personhood*

❧ General Limitations ❧

Absence of Enabling Conditions

Earlier I have described the *enabling* conditions of personhood. I have found existence in the mode of persons superior to other modes because it involves a genuine advance in my search for goodness. Persons acquire the values of individuality, moral integrity and autonomy. They find their surroundings suffused with meaning, which, when organized and unified in systematic ways—through science and philosophy—can induce in them highly satisfying feelings of piety and mystical oneness with all of my reality. The consequent or concomitant mood of such personal achievement comes in various shades of happiness and blessedness, in which human beings, while celebrating their own personhood, celebrate me as well. That's why the most important objective at the present stage of my career is to make persons *count*. In saying this I am reiterating my previously made claim that upon reaching the level of personhood I have found a means of making judgments about the meaning and value of what I am capable of. Persons monitor my well-being, and through them, in individual experiences, I really come into my own. Each person reflects me in a unique, irreplaceable way.

Humanity is diminished when its representatives, for whatever reasons, are deprived of the opportunity to acquire the enabling characteristics of personhood. Among the reasons may be unfortunate physical conditions, lack of essential resources such as food or fuel, inhospitable climate, famines, droughts, plagues, and epidemics, or invasions by hostile neighbors. While life is dominated by a desperate scramble for basic needs, the premium is on group survival, on herd instinct, on an absolute submission and total obedience to those who are strong enough to impose their will on the group.

When physical conditions of life become easy enough for cultural values to establish themselves, much study, observation, hard work, and thought is required to develop sufficient understanding of nature and human nature to organize life around values intrinsically worth pursuing. When requisite knowledge is not attained or is limited to only a small segment of the population, ignorance of important factors may prevent most individuals from acquiring capacities which allow personhood to flourish. But even in societies in which access to objective knowledge and to various other cultural goods is possible, unfavorable economic, social, and political arrangements may keep individuals from reaching their full personal stature.

Since the full flowering of personhood is always an open-ended ideal, never completely realized, the degrees of success vary considerably from society to society and from person to person. But it is an undeniable fact that by now humanity has reached a stage in which the appreciation of personhood is almost universal. The entire sweep of all civilizations in the course of human history has contributed to this happy outcome, in which practical invention, science, technology, religion, political and fine arts played an important role. Because the expectation of manifesting personhood is so widespread, it has become possible to regard dignity and personal rights as inalienable possessions of all human beings. Consequently, people are becoming more sensitive to situations in which some or other of the enabling conditions of personhood are not respected or are willfully denied.

The Drag of Tribalism

In many contemporary societies humanity has moved far beyond the stage in which the group's needs and goals dictated the behavior of individuals, who in fact were mere instruments of the group's will. With the growth of individual self-consciousness the standards and patterns of a group came under closer scrutiny by individual members, thus making possible a more knowledgeable and critical evaluation of the group's objectives. When distilled through perceptions and judgments of individual persons, common practices led to the refinement and flourishing of cultures, bringing into being rational ways of testing and evaluating the developing forms of life. Through consultation, joint deliberation, and cooperation among persons social groups could now be governed by explicitly formulated and rationally defended standards instead of blindly following rigidly laid down dictates, handed on with hardly any modification from generation to generation. Through original activities of persons various forms of art, religion, literature, and science gradually defined themselves. The very idea of progress, of improving practical skills and theoretical understanding, arose out of the awareness of improvements that could be introduced into a culture by individual efforts.

The process of change, infused with the momentum toward further improvement, is still going on in my human experiment. Unfortunately, in too many instances the optimum well-being of persons is blocked through subtler forms of dominance of the group over the individual. This blockage can be in part attributed to vestiges of various forms of tribalism which sacrifices the fulfillment of persons to the well-being of the group. I do not deny that many admirable values are social in character, such as the virtues of altruism or solidarity, for example. Good will, love, sense of belonging, patriotism are some of the finest emotions I celebrate in human experience. It is unfortunately also the case, however, that supposedly in the name of furthering these positive values personal life is too frequently diminished and even destroyed.

The task facing me today is to become more self-conscious—through individual persons, of course—of the ways in which social aspects of life inhibit or destroy personal well-being instead of enhancing it. An important part of this task is to show that people need not accept some limitations to their personhood imposed on them by such factors as sex, race, nationality, or ideology. This is not to say that membership in a group, or being of male or female gender, cannot supply a person with positive values. Indeed, I have just admitted that the sense of belonging, of participating in common cooperative ventures, are among the most important sources of blessedness. What is to be avoided, however, are practices which curtail or diminish the full personhood of individuals in the name of group membership. To circumvent this danger it is necessary to give persons an opportunity to take a stand on the value of their membership in a group and to decide for themselves what meaning this membership is to have for their personal lives. This opportunity is a necessary condition for developing autonomy and independence of judgment and action. The possibility of attaining autonomy is undermined by such dubious phenomena as the general doctrine of social determinism, certain interpretations of the sex differential, of racial differences, and of the role of political ideology.

☙ The Trap of Social Determinism ☙

"Society": A Weasel Word

One obstacle to developing autonomy is the doctrine of social determinism. That doctrine is invoked to support the claim that the beliefs and actions of individuals are determined by their social setting. Although frequently embraced, this doctrine is by no means clear. It allows a variety of distinct formulations. It may mean that the values and norms of society are reflected in the behavior of its members. Or it may mean that individuals *get* their values from the society to which they belong. Finally, it could be put in strongest terms, namely, that society fully *determines* the behavior of individuals. By this last view individuals are viewed as mere pawns in the hands of massive, irresistible social forces; human actions are seen as merely *reactions*, fully determined by the sum total of social influences exerted on the individual.

The first two formulations seem harmless enough and in a certain sense appear quite true. Without the heritage of tradition and culture from which to get a stock of knowledge and rules of behavior, without a multiplicity of practical and moral standards, a society could not even exist. But the third, the strongest version of social determinism, is problematic. Just how enlightening is it to say that individuals are products of their society? A meaningful answer to this question is precluded by the protean character of the word "society."

"Society" is extremely difficult to define; this term leads an adventurous life, the life of a weasel. Like the word "culture," it is used in ever new and changing contexts. The concept of society, or, more specifically, what each person understands by his society, is nothing constant. People relate themselves to their society in a way which depends on the content this notion has for them.

One may start with a rough definition of society as a group of individuals inhabiting a common physical space and sharing some aims, purposes, and regulative standards. As soon as one lays down some such definition an interesting fact obtrudes itself. Individuals are members not of *one* society but of numerous societies at the same time. Even as a child an individual finds himself moving in social groups that are governed by different norms and rules. There is the immediate family group characterized by intimacy, informality, close cooperation, constant companionship. From this the groups of family relations — aunts, uncles, cousins, grandparents — will be easily distinguished; what one can afford doing at home one will refrain from doing in the house of grandparents (or *vice versa).* When the child enters a kindergarten he will soon notice that it has an atmosphere, ethos, activities, and rules of its own. This situation will be even truer upon his entering a full-fledged school, which imposes new requirements and new regulations, at the same time enlarging the field of child activities.

Within the school society — other students and teachers — soon emerge smaller intergroups: classes, hobby groups, athletic teams, extra-curricular organizations. At the same time the young individual may become a part of a church community, again with its own set of activities and standards of behavior. As a member of the family one is likely to participate in social circles to which the family belongs; it will depend on the profession, education, interests of the parents whether one attends an occasional church social, a music concert, or a debutante ball. Upon entering a university the young person's membership in various social groups begins to proliferate even more. One will be at the same time a member of a class, of a college, of scholastic and social organizations, fraternities, of an athletic or political group. Termination of formal education brings with it the need for further, perhaps more important, associations. One will become a full-fledged citizen of a community, a state, a nation; one may even feel some loyalty to the world community as a whole.

The above reflections show how unilluminating it is to speak of *the* society to which individuals belong. Furthermore, membership in most social groups is voluntary. To choose a profession is to determine one's social connections. In a less pervasive way, persons determine what the nature of their society is to be when they join an athletic club, an art circle, or a political organization.

Socialization: Not Just Conditioning

It is important to emphasize that socialization of a child is not mere conditioning. From the very start, the inculcation of society's values addresses itself to the child's ability to reason. Parents and teachers presume that at a certain point the child will understand that there are good reasons for accepting these standards, that they are justifiable. Undoubtedly, the process of maturation is gradual. In early childhood, this inculcation of society's values proceeds by invoking the parents' authority, but even then there is a concomitant attempt to communicate to the child that the authority deserves credence because it is competent and always aims at securing the child's good. With the onset of adolescence, when the powers of independent judgment get strengthened, the appeal to authority is replaced by an appeal to the young person's ability to appreciate the values at stake and to commit himself to them out of his own conviction. In all but the most primitive, herd-like stages of humanity common social and moral bonds come into existence through such voluntary and reasoned commitment.

In a sufficiently advanced and diversified society, the identification of an individual with given groups will differ in seriousness and intensity. Participation in smaller and tighter memberships is likely to be more vital and to enter more effectively into a person's life. Conversely, a person can have a more definite effect on a small group rather than on a large one, such as a nation, for instance.

Community of Interpretation

It is important to point out, as did Josiah Royce, that every society to which one belongs is a *community of interpretation*. It is what an individual thinks it is. A university, for example, has for its members students, teachers, and administrators, and the way in which these members will view the purpose and the proper function of the institution will not be uniform. Within a political party there are likely to be "wings" — right and left, conservative and liberal. Even the participation in such a wide community as a nation is likely to involve tensions of interpretation. What is the right course for the nation to take: to join a political alliance or to remain neutral, to pursue a policy of isolationism or to operate on a basis of world community? Is a labor union to press for increasing improvements within its own ranks, or is it also to look to the benefit of the company and even the national economy as a whole? The latter can become a very live option during an economic recession.

Because the concept of society leads such an adventurous life the doctrine of social determinism is dubious indeed. It creates the impression that there is one Society in which individuals move, live, and have their being, while in reality every person is a member of a cluster of societies which to be sure overlaps at some points but in many respects is quite different. It is, therefore, unenlightening, misleading, and in some ways obviously false to say that people are merely products of society. A person who invokes the doctrine of social determinism in effect encourages an escape from the task of forging a personal autonomy. This task involves the active and ongoing process of interpreting the standards by which a given social group lives.

The need for interpretation makes each active member of a society a potential *reconstructive* center of that society. As long as society itself is developing, it can never be a sufficient determinant of the behavior of its members. Some groups do have greater influence on the behavior of individuals than others. But none of them can claim exclusive power over the judgment of individuals. One may take a narrow view and think of oneself only in terms of concrete relatedness to family, friends, community. Or one may also have a sense of belonging to larger entities, such as nation, mankind, or even all mankind in the past, the present, and the future. There is an infinite gradation of ranges which this relatedness to others may take on, depending on a person's education, general awareness and personal outlook. Undoubtedly, the political and economic structures of some societies exert a great deal of control over their members, thus affecting the limits of their individual autonomy. But to what degree they do so is a matter for investigation. As a general theory, the doctrine of social determinism is useless because it simply begs the question.

✂ The Sex Differential ✂

The Myth of Aristophanes

That people are born male or female is a fact of nature. Nature, however, does not tell them what they are to make of this fact. People may even wonder whether this is the best arrangement that nature could come up with, and they may speculate about other possibilities. I am intrigued, for instance, by an ancient Greek myth reported by Aristophanes in Plato's *Symposium*. According to that myth, there was a time when human beings were not separated into men and women. They were physically androgynous. "The sexes were not two as they are now, but originally three in number; there was man, woman, and the union of the two." That union was also physical—human beings were spherical, with four hands and four feet, and one head with two faces. Various advantages resulted from this arrangement, including the ability to move very fast, rolling like a ball. Being powerful and efficient, the androgynes, however, behaved in ways insulting to the gods and were punished for their offenses by being split in two, one half being made male, the other female. Now in order to perpetuate the race, they had to breed by mutual embrace, and this explains the sexual urge. "So ancient is the desire of one another which is implanted in us, reuniting our original nature, making one of two, and healing the state of man. Each of us when separated, having one side only, like a flat fish, is but the indenture of a man, and he is always looking for his other half."

This Greek myth explains, in an anachronistic sort of way, a syndrome expressed in the French motto: *Cherchez la femme!* One fact about the sex differential stands out. It happens to be my way of making sure that the human species continues; sex ensures procreation. The importance of that objective translates into certain biological and psychological factors. People rightly speak of sex as a basic need. Originating in a long evolutionary process, this need is as imperious as any instinct. Sexual interest indeed arose out of instinctive forces, and is a form of *conatus*. Its power is recognized by people when they use such locutions as sex *impulse,* sex *drive,* or sex *urge.*

Another fact emerges from my—that is—nature's sexual arrangement for the continuation of the species. For some reason that fact tends to be overlooked, despite, or because of, its obviousness. Both men and women have a stake in the objective of the sex differential. The fact that the perpetuation of the species depends on the collaboration of the sexes is of crucial importance; it is the original and primordial prototype of *community.* To the extent that men and women care about the continuance of life, they are committed to a common objective. Thus the sex differential is at the same time the ground for the emergence of mutuality and the acknowledgement of interdependence.

Sex and Culture

The presence of any differential, no matter how important its basic purposes, can spell trouble. The very perception of differences can raise the question of their relative importance. The ancient battleground of the sexes is the interpretation of this difference. Nature does not dictate *how* the sex need is to be satisfied. Sexuality is complex and involves courtship, intercourse, and pregnancy with its equally complex consequences; it affects life in numerous ways. In this natural complexity lies the opportunity to devise a great variety of standards and norms of behavior, each dealing with some aspect of human sexuality. Students of human cultures have indeed uncovered such a variety in actual practices. Among other things, the sex differential has given rise to a very broad notion of masculine and feminine *roles.* Often these roles were interpreted all-inclusively and were understood hierarchically, with either women or men playing an overall dominant role. In societies where such dominance is not explicit or not general, men and women may be still assigned definite roles, the performance of which conditions their respective life patterns. Men are expected to play the roles of providers and protectors; women are charged with domestic chores and child-rearing.

The presence of a difference naturally invites comparison. The result of that comparison may be a desire to dominate, based on an evaluation of the meaning of that difference. Such an evaluation in terms of functions or roles is inherently dangerous, even though it can facilitate certain desirable social results. Plato's *Republic* is notorious for its advocacy of a division of functions in society. The utopian state, according to Plato, would delegate the performance of specific tasks to special, designated groups, membership in which is to be determined by abilities and special training. The *Republic* advocated a class society, consisting of guardians, soldiers, and artisans, each class being entrusted with the exclusive performance of certain tasks. Along with responsibilities went, of course, certain rights and privileges. The resulting inequalities Plato justified by the successful functioning of the state. The end justifies the means.

If one looks at social arrangements across the span of human history, one will discover a great variety of ways in which, in the quest of certain desired results, groups of human beings were treated primarily as performers of certain roles or functions. The actual result was almost always a curtailment of other inherent capacities and needs. In that sense, human history is a history of a subjugation and dominance. Difference in social function invariably tends to result in inequality of treatment, in curtailment of rights and privileges. This tendency is clearly evident in the way men treated women. Although Plato thought that women's capacities were not sufficiently different from those of men to deny them the right to be trained as rulers, he, like Aristotle, and the Greek culture as a whole, regarded women as inferior to men. All Western history, up to the present, shows the marks of discrimination and sexual prejudice against women. Examples are not hard to come by. Countless texts can be cited to show that the Western tradition tends to deny to women equal moral status, and that it does so on the basis of mere sex differential. Even such a balanced philosopher as Aristotle was prone to make judgments that are grossly prejudiced:

> ... it is a general law that there should be naturally ruling elements and elements naturally ruled... the rule of the free man over the slave is one kind of rule; that of the male over the female another... the slave is entirely without the faculty of deliberation; the female indeed possesses it, but in a form which remains inconclusive.[11]

The prejudiced sexist view can be found in religious scriptures as well. "How can he be clean that is born of a woman" (Job, 4,4). The Daily Orthodox Jewish Prayer (for a male) includes this: "I thank thee, O Lord, that thou has not created me a woman."

Saint Paul wrote: "Let the woman learn in silence with all subjection. But I suffer not a woman to teach, nor to usurp authority over the man, but to be in silence." In Ephesians 5:23, we read: "Wives, submit yourself unto your husbands... for the husband is the head of the wife, even as Christ is the head of the Church.''

How such one-sided, prejudiced, and unfair cultural generalizations about women came to be generated and on what experiences they were based is not easy to determine. However they got started, they certainly had a snowball effect. One-sided views tend to evoke one-sided response, thus increasing the gap of mutual misunderstanding. Being put at a disadvantage cannot remain unnoticed forever, even if a reaction may be slow in coming. When opportunities for personal development and self-expression are denied to women, it should not be surprising that qualities dependent on such opportunities are not present. This applies also to the ability to perceive one's own plight. If only few women get an extensive education, few of them can do things that presuppose such an education.

Many characteristics ascribed to women are due to frustrations resulting from the very fact that they were *encouraged* to see themselves as having a dependent, subordinate status. If women are seen as primarily sexual creatures, discharging mainly a biological function, then it is no surprise that they, in self-defense, may be tempted to resort to using that feature in order to gain some ascendency. Feminine ploys or power plays arise from the frustration of not being allowed *to act out their own humanity* and being forced instead into playing up and exaggerating their femininity. In other words, they are driven to engage in something like a generalized, even if etherealized, form of prostitution. Since men are not invulnerable in sexual matters, and, like their counterparts, suffer emotional and psychological frustration when sexual life is unsatisfactory, they may be prone to exaggerate the destructive female powers. This phenomenon seems to account for the frequent characterization of women as seductive (Eve *tempted* Adam), irrational, an amoral or even anti-moral force. Folklore and literature abound in examples of presenting woman as an elemental force which dissolves discipline and induces abandon. Woman is seen as a vampire, depriving man of his vitality and inner peace. She is the disrupting, disturbing element, the Earth Spirit in its sinister aspect.

Sex and Love

That sexuality is a powerful force is not a secret to either of the sexes. R.M. Rilke referred to it in his *Duino Elegies* as Neptune, the river-god in the blood, thus pointing out that the underpinning of romantic love is the ancient biological reproductive force. That ancient force of life in the human race lends irresistible intensity to individual loves. ''We do not love like flowers, with only one season behind us. Immemorial sap rises in our arms when we love,'' says one line of Rilke's poem. But Rilke does not limit himself to calling attention to biological roots. He also shows that human beings are capable of disarming the threat of this mysterious life-force by domesticating it, by dissolving its pressure in gentle, caring, compassionate, and understanding behavior. By understanding the origin and the effect of the sexual impulse persons can guide it along avenues that preserve excitement and fascination but also allow them to transform it into mutual affection and comprehensive, not merely physical, love. The poet describes how the love of a mother calms the turbulence of her child's sexually-induced nightmare, and how a lover can act as the quieting, anxiety-countervailing agent. Love between persons is a libido that is morally and esthetically transformed.

Rilke's perception of the multifacetedness of human sexuality calls attention to the fact that its form and its meaning depend on interpretation. Traits associated with femininity or masculinity are *culturally* determined. Even the notorious "macho" conception of masculinity is not inevitable and is absent from some cultures. Consider, for example, the etymological root of the word "gentlemen." In the sixteenth century Europe a gentleman was gentle. He played the lute and sang; he was appalled when he discovered a callus on his hand. A culture may instill certain attitudes and dispositions when it encourages the boys to play with toy soldiers and girls with dolls. Some psychological traits and dispositions do not have their source in biology but are the result of conditioning, sometimes reflecting the disappointments brought on by certain practices of the dominant sex. When women are seen as frail, fickle, unstable, or unfaithful, the most likely background of such opinions is a social arrangement in which absolute submission to the male will be expected. Some interesting consequences of this syndrome have been explored in literature.

Women As Moral Teachers

Friedrich Hebbel's play, *Herod and Marianne,* poignantly presents the effect of a despotic male distrust of women. Before leaving his court on a dangerous mission, Herod demands from his beautiful wife Marianne that she promise to kill herself if he does not return. The playwright makes it clear to the audience that she fully intends to do so, but at the same time she expects her husband to trust her loyalty and therefore refuses to promise that she will commit suicide. Herod does not trust his wife and arranges for someone to murder her if she fails to kill herself upon learning of his death. When he makes this arrangement twice in a row, she teaches him a lesson by pretending that she was glad of his supposed death and was marrying another. He has her executed, only to learn belatedly the truth about her undying loyalty. Thus Marianne punishes Herod for deeming her incapable of loyalty, and in going to her death she affirms her inviolable dignity as a person.

Another rare example of crediting women with equal and even superior intellectual and moral powers is found in Goethe's play, *Iphigenia in Tauris*. In this adaptation of an ancient Greek play, Goethe presents Iphigenia as a noble and courageous defender of human dignity. Banned by the gods to an island for the sins of her parents, Iphigenia is put in charge of a sacred grove by the barbarian king Thoas. Discovering that the barbarians practice human sacrifice, she persuades Thoas to abandon the practice. As he gets to know her better, Thoas, impressed by her beauty and intelligence, wants to marry her. Iphigenia's hope, however, is to return to her native Greece, and she puts off the decision. Unexpectedly, her brother Orestes arrives secretly on the island with a plan to rescue her. At first she agrees to participate in the rescue scheme. But then she has second thoughts. She hesitates to betray Thoas's trust in her by stealing away in secret. In the course of her stay on the island she had established with him a deep human contact, one good result of which was the abolition of human sacrifice. Moreover, even though she is his captive, he treats her with dignity and offers his hand in marriage.

Although strangers and even enemies, the Greek priestess and the barbarian king have developed a strong moral bond. By escaping, Iphigenia would destroy that bond. So she decides to put this bond to a test. She reveals to Thoas the secret plan for her abduction. The potential consequences of this disclosure are disastrous. Learning of the presence of hostile Greeks on the island, Thoas can destroy them, including Orestes, Iphigenia's brother. But she tells him to his face: Destroy us, if you dare. In doing so, she challenges his moral values. She describes in glowing terms the positive effects of Thoas voluntarily allowing her to return with her brother to Greece. The result of such a free decision on his part would be not only a deepening of their mutual respect but also an initiation of a friendly relationship between the Greeks and the barbarians who are now hostile to one another. Iphigenia's appeal to Thoas's moral sensitivity works. He allows them to go, thus ushering in a new kind of relationship between individuals and peoples.

Goethe's message in this play is that the standard of universal respect for human dignity transcends sex. He communicated a similar point in his great drama *Faust*, in which the innocent young girl Gretchen is presented as eligible to instruct Faust, a titanic male, in the ultimate wisdom of life. As Faust's guide to final salvation, she illustrates the meaning of Goethe's phrase, "The eternally feminine draws us upward."

These examples from literature recognize the moral equality of persons regardless of their sex. The relation between the sexes needs to be subordinated to their relation to one another as persons. Men and women as persons can recognize mutual dependence and establish a primeval sense of community. This sense of community and its intrinsic benefit to persons can be best perceived in the love of offspring. It is natural for people to care for their children, no matter whether they are boys or girls. Children are loved as human persons, regardless of sex, even though for various economic and social reasons in some cultures offspring of one sex are preferred over the other. It is important to keep in mind this ultimate status of human beings as persons, because it is only by reference to this basic equality that any differential, including the sex differential, can be managed in a rational manner.

Personal Equality of Genders

Like other artificially-created and blindly-perpetuated inequalities, the sex differential can be transcended when persons become aware of and take seriously the equal dignity of every human being. The claim to dignity rests on every person's *capacity to judge* whether the invoked standards and roles are beneficial to both sexes. If to embrace a certain standard entails an important loss to either sex, then that standard is questionable. With regard to the ability to judge such standards, there is *no* difference between men and women. Whatever plausibility the ascriptions of some temperamental differences have, the plausibility is due to the conditioning men and women receive as a result of having been assigned by their culture certain roles and functions, and having been brought up by the tradition to see themselves in certain ways.

Seen against the background of community and mutuality, the sex differential is a definite boon. So-called romantic love is a response of the two human forms to each other. The very fact of difference between men and women is a positive plus for life. As the myth invoked by Aristophanes vividly illustrates, the sexually-conditioned differences have the attraction of complementariness. The physical appearance of human bodies has formal, esthetic features. These features are also found in behavioral expressions, such as grace and charm in demeanor and bearing. There is no reason why to each sex there should not be corresponding types of such expressions. That there are different masculine *and* feminine esthetic features is a gain, not a loss; uniformity would be boring, and probably not as effective biologically. The French have a point when they exclaim: *"Vive la différence!"*

The sex differential provides for a special type of complementarity of human excellences. A good marriage thrives on this complementarity. Besides the esthetic attraction, it can include a mutual admiration of traits that harmonize with partial roles voluntarily played by each partner. There is no reason to deny that acting as a lover or as a mother is enhanced by allowing gentleness and softness to prevail, or that having been endowed with a heavier, more muscular body allows the male to act resolutely and protectively on appropriate occasions. A distinction between typically feminine and masculine virtues can be made without thereby lapsing into invidious sexist discriminations. Such a distinction does not deny that men can be gentle, or women firm and that they can appreciate these qualities in one another. The essential objective of interpersonal relationships should be to elicit what is best in each person as a human being and not as a representative of a particular gender. No one likes to be reduced to the role he or she plays. As Sartre rightly claimed, it is logically impossible for a human being to bring off a full and exclusive identification with a role. The harder one tries to play the role of, say, a waiter, the more obvious it becomes that one is *not* that role. The same applies to the sex differential. The harder one tries to play at being nothing more than a female, or a male, the clearer it becomes that to play that role a person has to transcend that role. This is due to the fact that men and women are primarily human beings, not confinable to any role. There are many ways of being a man or woman. *First of all and ultimately, people are individual persons.* At the level of irreducible human personhood, people can build relationships that engage their capacities as persons and can construct communities that reflect their mutually arrived at values. Equal rights for men and women are possible because at the level of common humanity the differences disappear. These rights are not just comparable but identical, due to them not in virtue of performing certain roles but simply by virtue of being persons.

❧ The Ethnic Mask ☙

Ethnicity and Culture

Like the sex differential, the racial or ethnic identity is something given. Being born male or female is a dispensation of life; being born into a particular ethnic group is a matter of historical accident. Humanity invades persons in some particular form: male or female, black, brown, or white, French, Russian, English, or Chinese. True, national or religious identity is something that can be changed or discarded, in contrast to the color of skin or the sexual physiology. Recently, however, even the latter has become subject to surgical manipulation. Similarly, attempts to find out whether the sex of offspring can be controlled have generated some theories as to how it might be done. Nevertheless, should these attempts prove successful, the determination of the unborn child's sex will still be in the hands of the parents. Instead of being the gift of nature, one's sex will be due to parental decision, and if one doesn't like one's sex, one will be able to blame one's parents.

In any case, all persons face the task of working with the initial data of human reality. But what is done with these data is not predetermined. It depends on how natural endowment is perceived; the meaning of the situation into which people are born is a function of interpretation. (Because of different interpretations, sexual roles and their relative importance can vary from culture to culture.) The same is true with regard to other inherited characteristics. If humankind were only of one race and had not differentiated itself into tribal and national groups, some problems would not arise. But as the story of Babel illustrates, the mere presence of linguistic division can result in a lack of mutual understanding and even in hostility among human groups.

Inhabiting many continents, facing different geographic and climatic conditions, and being separated for long periods of time by natural barriers, human groups have devised a great variety of ways of relating themselves to nature and to each other. In a global perspective, the richness of cultural traditions created by racial and ethnic groups makes my life in human form colorful and varied. From this point of view, I take delight in the fact that in the house of humanity there are many ethnic mansions. A student of world civilizations is bound to be impressed by the fertility and the ingenuity of the human spirit. It accommodates itself to varying conditions and interprets them in the light of myth and theory. It celebrates experience in music, poetry, drama, literature, and other forms of art. It creates, transmits, and utilizes knowledge of various kinds. Effective or flawed political arrangements of human history unfold the drama of the rise and fall of empires, the upsurge and decay of industry and commerce, the flourishing of science and philosophy, art and religion. Although human groups often managed to accommodate themselves to others, mutual hostilities and wars thrive on the perception of those others as radically different.

Ethnocentrism v. Universalism

It seems natural to start with the belief that the only way to be is one's own, and that all other ways to be are deviant—others are strangers, heathens, or barbarians. This tendency is reinforced when the others repay the compliment with equal contempt for and hostility toward one's way of life. In addition, the competition for scarce resources compounds the mutual estrangement. If one lacks the land for grazing or agriculture, it seems natural to take it away from one's enemies. Here human behavior follows patterns established in pre-human forms. For a lion it is neither moral nor immoral to kill a deer for food. In a human group the ground of hostility may be a perception that others live by incomprehensible beliefs and follow reprehensible practices. The conquests and reconquests of the neighbors' land or property can be laid to this elementary fact.

There were, of course, contacts and relations that were peaceful and friendly in nature. These gave rise to trade, commerce, and exchange of knowledge. Intensive and prolonged exchanges contributed to the emergence of civilizations, some of which are credited with hastening moral progress. Thanks to peaceful explorations, expeditions, and alliances, mankind has achieved a modicum of global community. At first predominantly satisfied with local deities, with the "God is with us" syndrome, religion gradually rose to the more lofty conception of *one* God who is equally concerned about every person and every nation in the world. Monotheism was the victory of the principle of equal love and uniform justice for all human beings. It made possible more objective evaluation of perceived differences between human groups and invited a scrutiny of their respective moral standards and practices. The universalistic thrust of religions, proclaiming all human beings to be God's children, placed in question the assumptions that tribal and national group loyalties are the highest loyalties one can generate. Having first surfaced in the thought of some morally sensitive individuals, the desire to eliminate irrelevant and invidious distinctions affected not only the world religions but also the modern secular proclamations of the brotherhood of all men, as voiced by the French and the American revolutions of the eighteenth century.

This universalistic thrust is fuelled by the same insight which made it possible to perceive in the sex differential a potentially restrictive and discriminating barrier. No less than the discrimination based on sex, that based on being born into a particular racial or ethnic group may be as invidious and unfair. Consider an example. The separation of the races has tragically affected the history of the United States. Besides leading to a fratricidal civil war, it has left a legacy of hostility, discrimination and misunderstanding. As the result of the recently won civil rights legislation, inspired by courageous consciousness-raising efforts of such leaders as Martin Luther King, Jr., the Black minority has shaken off to a large degree the sense of inferiority forced or foisted on it by the dominant White majority. The slogan "Black is beautiful" has translated itself into restoration of self-pride and celebration of one's equal worth and dignity as a human being. Social and economic changes, although initiated by the force of law, have gradually lifted at least some of the psychological burden of being treated as a member of a second-class minority.

The lifting of this burden was made easier by the consciousness of having preserved one's fundamental human dignity under the adverse condition of slavery and social repression. Even under these conditions, the Blacks reaffirmed their own worth by exhibiting such admirable moral characteristics as resilience, solidarity with one's kind, equanimity, and even cheerfulness and good humor in the face of adversity. As many studies have shown, underneath the appearance of equanimity there often seethed the destructive psychological pressure of repressed rage. Considering the burden of that pressure, it is impossible not to admire in the behavior of representative members the almost Stoic dignity and noble resignation of the oppressed race.

The experience of the Blacks in America is an instance of a more general phenomenon observable in the life of any minority group, whether it is racial, national, or ethnic. Minority groups tend to develop from within their human resources certain special abilities and characteristics that gradually become incorporated in their traditions and self-consciousness. Under adverse and prosperous conditions human groups tend to generate some unique attitudes and modes of behavior. Not without reason do racial, ethnic, national, or religious groups take pride in some of their special traits or excellences. Such excellences not only help them to survive under hostile conditions, but are also cherished and celebrated as special expressions of the group's spirit. In contemporary Black American culture, the new sense of confidence and pride is no doubt inspired by honoring its special and unique contributions.

But there seems to be a danger in tying one's self-worth to some specialized distinctions, a danger which exists for *any* ethnic group. If one's identity as a human being is exclusively expressed in a particular ethnic garb, then one is going to be constantly faced with the task of comparing one's ethnic traits with those of outsiders. Such comparisons are likely to fill one with anxiety: am I better than they, or are they better than me? One will look for features and characteristics that are *different* from those of others, and may often be tempted to conclude that these characteristics are better in virtue of simply being different.

The process of invidious comparison is likely to be reinforced by the memories of injustice and maltreatment suffered at the hands of the majority. It may perpetuate a hidden inner prison of siege mentality, requiring one to be constantly on guard. For instance, the need to keep proving to oneself and to others that Black *must* be beautiful can become a psychological burden. Not unnaturally, it can evoke a counter desire in those from whom one expects recognition and respect, namely, a desire to prove that White is beautiful too. If both Blacks and Whites are thoughtless enough to embark upon this silly game, they will only succeed in making themselves nervous in each other's presences. It is also likely that in these exchanges those who in the past enjoyed the dominant position will seek, in part as a defense mechanism, to justify their unjustifiable former ways by unfounded and unsupportable claims, racial, religious, or nationalistic, thus perpetuating mutual alienation and suspicion. This story is an old one, tragically enacted in all parts of the world throughout human history.

The Common Humanity of Persons

It should not be denied that a part of a person's identity *is* culture-bound, and the richer that part is, the richer the person. But ethnicity is not enough. One must be able to establish one's *personal* identity *vis-a-vis* others, no matter what one's ethnic background is. This need is especially urgent when the members of diverse human groups are inevitably in constant interaction. They cannot genuinely interrelate in terms of diverse racial, ethnic, or national values alone. To understand each other — in the inclusive sense of that word — they need a common bond. Where can such a bond be found?

The answer to this question is straightforward: *in their common humanity.* In addition to having an ethnic background all persons share a more inclusive human heritage. That heritage makes it possible for them to transform their particular ethnic background into a *personal* outlook in which are incorporated some common standards vital to inter-ethnic communication and harmony. Among such standards are mutual respect, acknowledgement of complementary excellences of others, helpfulnesss, tolerance, and good will. No ethnic group has a monopoly on these virtues. In this sense, neither Black nor Brown nor White but *Human* is beautiful. The really rewarding relationships between members of diverse ethnic groups come into being when they begin to develop bonds that move beyond (or behind) ethnic values and are based on personal traits and interests. This realm is where enduring and meaningful integration occurs. Everything short of such a personal respect is integration only in name and appearance, not in reality.

This statement is not to deny that among the values out of which one forges one's own personal destiny are particular cultural and ethnic values. Obviously, they are the natural starting point. One is born into them, and quickly they become second nature. But if one is in contact with other ethnic values, as is inevitable in a multi-ethnic society, one cannot help being conscious of the possibility of alternative values and styles. These styles and values are not always live options. But one may come to see that, if these options are closed, they are closed not because of some iron necessity, but simply because of the accident of being born into a certain ethnic, social, or economic group. *Humanly* speaking, a Black can have the lifestyle of a White and vice-versa. This truth is especially transparent for offspring of racially mixed marriages. "But for the half of my parental origin, I could be Black *or* White"—this is a perfectly valid observation for many persons.

I wish to emphasize my basic claim that the attainment of all virtues and excellences is always personal. They cannot be instilled by mere conditioning, or social engineering, or brainwashing. Like arithmetic or physics, they must be acquired by individuals and expressed in personal behavior. All learning is individual and often a lonely process; it does not happen as a result of mere togetherness or gregariousness. It certainly does not happen by osmosis. Homework is one's own, not someone else's. Similarly, a character trait is one's own and is not automatically bestowed by a membership in a group. No matter what one's sex or ethnic background, one must respect oneself primarily as a person.

The task of becoming a stronger and better person is universal and endless. It is faced by members of minorities and majorities, by men and women, by agnostics and religious believers. If one derives a sense of self-importance from *merely* being a member of a group, one is making a claim to which one is not entitled. Mere membership in a minority or a majority is not something achieved but inherited and as such confers no merit. In contrast, a skill, a proficiency, or a virtue is attained and maintained by personal effort. It requires constant attention and recommitment.

A society is in danger of becoming stagnant if its members expect everything good to come about by impersonal economic and governmental institutions and it must be reminded of the importance of the individual, personal dimension. This reminder is not a call to put all the burden back on the isolated and powerless individual, thus distracting his attention from the importance of social and political action. The need for organized social and political action is undeniable and unquestionable. But that action can be intelligent and effective *only if* it emanates from strong, informed, and confident individuals. This was the conclusion of a writer who, himself a victim of an oppressive political regime, pondered deeply how a society can overcome its own corruption. Consider the view expressed in Alexander Solzhenitsyn's novel, *The First Circle*. The novel's hero, the engineer Nerzhin, political prisoner about to be shipped to the camps, says the following about those who proclaim the superior wisdom of "the people":

> What was lacking in most of them was that personal *point of view* which becomes more precious than life itself. There was only one thing left for Nerzhin to do—be himself...[He] understood the people in a new way, a way he had not read about anywhere: the people is not everyone who speaks our language, nor yet the elect marked by the fiery stamp of genius. Not by birth, not by the work of one's hands, not by the wings

of education is one elected into the people.
But by one's inner self.
Everyone forges his inner self year after year.
One must try to temper, to cut, to polish one's soul so as to become
a human being.
And thereby become a tiny particle of one's own people.[12]

Solzhenitsyn's absolute affirmation of the infinite worth of every person, regardless of sex, race, religion, or political beliefs, recognizes the relative and subordinate value of these differentials. Any of them, if absolutized, violates the dignity of human personhood. If they are not judged in terms of the over-arching standard of respect morally due to every person, they can result in tyranny, exploitation, and discrimination. On the other hand, if such respect is granted, persons' autonomous creative activities can enrich the fabric of social existence. As the recognition of sexual differences makes the relation between sexes much more interesting than it would be if people were all of one sex, so the presence of ethnic and national traditions adds color and variety to the life of humanity. People can be proud of their ethnic roots without forgetting that these roots provide only one source of potential flourishing as autonomous persons.

◈ Ideological Warfare ◈

Two Social Ideals

Another contemporary phenomenon which prevents countless persons from allowing their full personhood to emerge is the worldwide ideological conflict. The world is split into two antagonistic political camps. Ideologies are not inherently negative phenomena. Any set of ideals or moral convictions, if it is to play a significant role in human life, needs to be translated into action. When social ideals get translated into political programs, they become ideologies. Not limiting themselves to mere rhetoric or persuasion, they tend to accumulate social, political, and economic power. When the use of that power, however, becomes questionable, "ideology" acquires a negative connotation. But if by "ideology" is meant merely a set of ideals and corresponding institutions by which people live, then even a democracy as a viable social and political system is an ideology in the broad sense of the term.

The present division of the globe into two ideological camps is in part due to the fact that each is dominated by a different social ideal. Western democracies emphasize freedom, while the Marxism-oriented countries give primacy to equality. A sizeable portion of the globe, comprising nations based on traditional Western democracies, is anxious to preserve as many freedoms as possible, in the belief that only these can assure optimal economic productivity while fostering the overall cultural and moral well-being. Inequalities arising from the free interplay of social, economic, and cultural forces are seen as helping to bring about diversity and contrasts. These not only stimulate economic productivity but also release creative forces from which everyone benefits in the long run . The basic argument is that if the inequalities arising from this free interplay of productive forces were to be eliminated, those on the bottom rung of economic well-being would be worse off still.

Another sizeable portion of the globe is governed by political parties that want to achieve social justice by direct action. Inspired by the Marxist ideology, these groups are willing to impose on the people a system in which not only the economic but also all other institutions are centrally controlled and in which the production and distribution of goods — economic and cultural — is tightly supervised by the government. The freedom of initiative, of forming independent associations, of devising alternative styles of living, is suppressed, because a conformity of beliefs and behavior is seen as a necessary condition for efficient production and just distribution of goods.

Looking on and trying to decide their own course, there is the relatively uncommitted portion of the world population, politically sandwiched between East and West, Marxism and capitalism. The ideological contest between East and West is for the allegiance, the minds and hearts of those populations that so far stand on the sidelines, uncertain of what course is preferable. This hesitation is understandable because an impartial observer is likely to notice some negative consequences of concentrating exclusively either on the ideal of justice or on that of freedom.

Alexander Solzhenitsyn and countless other eyewitnesses have provided devastating first-hand accounts of the immense suffering imposed on millions of people in the name of justice and equality. The draconian measures taken by the Soviet government to eliminate opposition to its programs have appalled the world and have raised doubts in many people's minds whether *any* form of socialism can avoid totalitarian oppression and elimination of basic freedoms. It is the same Solzhenitsyn, however, who is deeply distressed by what he regards as abuses of freedom in the West: lax morals, permissiveness, disrespect for public institutions, ubiquity of shady and criminal schemes for the sake of a quick buck, vandalism, littering, low esthetic level of popular music, mindless entertainment, sensationalism in journalism, scant interest in serious literature, intrusiveness and pushiness of advertising. Critics of Solzhenitsyn point out that he does not understand the workings of a democratic society, suggesting that some unwelcome by-products of freedom are inevitable. Yet it is clear that the ubiquity of the same unlovely phenomena—the rise of violence and crime, and the proliferation of unethical conduct among public servants even on the highest level of government—bothers many friends of democracy who see the need to curtail some freedoms for the sake of the general well-being of society as a whole.

The Costs of Ideological Rigidity

The ideological struggle between the two camps of humanity is of concern to me for many reasons, the most important of which I discuss later in my story. But even apart from that ultimate danger to life on earth, I deeply regret the price for senseless ideological hostilities that is being paid by multitudes of persons. Here I see another example of the tendency of human groups to freeze their standards into rigid abstractions, a persistent error which humanity has yet to learn to avoid. That error has its origin in the distant past of the human race when it existed at the level of strict tribalism—where there was only one imperative: the will of the group. Although human beings can move toward a different, more advanced level by becoming critical of standards by which individual behavior is to be governed, vestiges of tribalism are cropping up. They cling to human behavior when they coerce a person to act only in the name of some special group—a gender, a race, a religion, a nation—and so prevent the expression of a personal judgment, or a critical appraisal of the group's values.

I am suggesting that modern exclusivist ideologies are vestiges of such tribalism, precluding flexibility and compromise which would save many a person from being destroyed by the rigid demands of ideological orthodoxy. Even a valid standard may become an empty abstraction, disconnected from actual facts. Thus the ideals of freedom and equality can be invoked to construct social and political structures which sacrifice countless individuals at the altar of the alleged common good.

To avoid such an abuse of ideals, I propose one simple test: how does a social or political program affect the overall well-being of *living persons*? To apply this test is to keep checking in what ways the enactment of an ideal connects with actual life. Exclusive concern with freedom can easily result in unjust practices. Similarly, a determination to bring about justice based on egalitarian principles notoriously leads to drastic curtailments of freedom. The danger of reifying an abstraction, thus disconnecting it from actuality, is common enough and all-too-human. One, therefore, should not forget that wisdom has no other habitation, no "ontological status," apart from the correct application of the adjective "wise" to some persons or acts. Similarly, there is no freedom where *people* are not free, and no justice when people do not deal justly with one another.

The one-sidedness generated by a devotion to one social ideal at the expense of another can be avoided if social ideals are evaluated in terms of their actual effect on persons. Those who appeal to the ideals of freedom and equality in establishing institutions and practices must keep asking themselves whether and to what extent concrete needs of persons are actually met.

Western democracies acknowledge this requirement, always in theory and often in practice. For example, the concern with the rights of persons to seek a good life through individual initiative lies at the heart of the American political system. One reason why becoming an American looks so attractive to potential and actual immigrants is the awareness that America is a country in which particularistic features, such as ethnic background or religious conviction, have no or minimal bearing on one's acceptability as a person or a citizen. This factor is no less important than the obvious attractiveness of economic opportunities. The ethnic variety of the population undoubtedly plays a role in preventing the emergence of parochial nationalism. Equally important is the principle of the separation of church and state; it ensures that in religious matters people are free to follow their intellectual consciences without ever worrying whether their beliefs are acceptable to their neighbors or co-workers. It is a liberating discovery to realize that one's worth as a person is not tied to the condition of being of "the right faith."

The arrangement which allows for a variety of privately-held beliefs to co-exist with an allegiance to a minimal set of political beliefs is based on the respect for persons as accountable on a public level and free in private life. The genius of American democracy rests in this double principle of responsible citizenship and personal freedom. The inherent reasonableness of this principle accounts for its universal appeal, and this is what *America as an idea* has to offer to the world. Indeed, the distinctive feature of American democracy is that its institutions manage to keep the double principle of public allegiance and personal freedom alive. This principle assures a built-in flexibility, an ability to introduce in an orderly constitutional and legal manner such changes as appear desirable in the light of shifting circumstances and growing knowledge. To perceive the need for such changes and to initiate them is always the responsibility and the task of persons who live by that principle, particularly those elected or appointed to positions of leadership. Which tasks need to be performed and how is always a question of judgment, to be exercised by persons participating in the system, which ultimately includes every citizen.

However, while insisting on the importance of preserving individual freedoms in their quest for justice, Western democracies need not scoff at developing societies that seem impatient to pursue the goals of justice and equality even if they entail extensive denials of personal freedoms. There is no need to question the moral validity of the professed objective to attain well-being for all persons, regardless of race, sex, nationality, or creed. Having made justice and equality its *primary* political norm, Marxism understandably has a tremendous appeal to those who have legitimate grievances in this regard. Presenting itself as an antiprivilege and pro-equality movement, communism looks attractive to those who suffer from conditions of inequality and privilege. Living under such conditions, they are not likely to be impressed by or even pay attention to the fact that the chronic difficulty for communist and socialist governments has been their inability to harness the motivational resources that would ensure a productive and yet unoppressive society.

To grant the moral validity of the ideal of equality does not mean, however, to acquiesce in the practices that violate human dignity and rights. There is another alternative. Critics of Marxism can persist in presenting arguments that vindicate the policy of encouraging maximum production and fair distribution of wealth to all citizens while maintaining the conditions of freedom. The experience of democratic societies indicates that the tapping of economic resources occurs best when human energies are optimally released and that a society governed by free parliamentary procedures is likely to pass and enforce laws that in the long run eliminate gross privileges and inequalities.

The Path Toward Accommodation

It would be unrealistic to expect that even a prolonged and genuine debate between the two ideological camps about the best ways of combining the ideals of freedom and justice will quickly lead to harmony and unanimity. But I should warn that the alternative to debate is growing mutual incomprehension and hostility, with military confrontation as a predictable result. While remaining economically and militarily strong, Western democracies must also stay patient and aim at long-run results, in the meantime welcoming various attempts in the Marxist world to diminish dogmatism and oppression.

History is not easily "rolled back." It is not a foregone conclusion, for instance, that the countries of Eastern Europe would have immediately abandoned socialism of any form had they been allowed by the Soviet Union to carry out their grass roots reforms—in Hungary, in Czechoslovakia, in Poland. With regard to the Soviet Union itself, even such a prominent and effective critic of the Soviet regime as Andrei

Sakharov does not advocate the scuttling of socialism in his country. Neither do most of the dissidents in other communist countries. The example of Yugoslavia shows that countries tend to follow the internal momentum of political forces and seek solutions to their social and economic problems by drawing on their own experience and native resources, while enlisting only limited external advice, guidance, and support. The example of China shows that an indigenously developing social and political revolution, even if it starts as a violent upheaval, can nevertheless take a form of ongoing self-revision and modification, prompted by objective weighing of the consequences of previous decisions.

By the same token, it is an open question whether the Western democracies would necessarily be threatened by the emergence of states that decide to experiment with some forms of socialism. Some such experiments, with mixed results, have and still are taking place in various European, Asian, African, and Latin-American countries. Given the oppressive feudal tradition of some countries, it is not surprising that those seeking to overcome it look to the socialist alternative as more feasible. Just what *is* feasible or viable is an open question. A peaceful and mutually beneficial coexistence of capitalist, socialist, and mixed systems is precluded only if the very possibility of the emergence of such coexistence is blocked by hostile diplomacy and military force.

One inescapable fact dictates the necessity for the East and the West to seek mutual accommodation, namely, the increasing global interdependence brought about by instant communication and easy mobility. The volume and the nature of international trade changes rapidly due to the emergence of new industrial capabilities and to the growing labor force. These changing patterns of production and trade call for new ways of balancing economic gains and losses. Some problems arise from primarily political decisions, in which it is all too easy to pursue self-interest at the expense of those whom one regards as enemies. Others are the result of strictly economic or business decisions that need to take into account the growing competence of the labor force in the world. Steadily increasing trade calls for mutually beneficial pricing and trading policies. Old patterns are almost imperceptibly but irrevocably upset, leaving no clear alternatives in sight. Some adjustments are painful, even though in the long run they may benefit all parties concerned.

The increasing mutual involvement is not just political and economic, it is also cultural. People all over the world can now look into each other's backyards. They not only buy material goods but also learn and borrow ideas from one another. Increasingly conscious of their own historical and cultural identity, the so-called Third World countries clamor to be included in the global conversation and are eager to take part in economic and political decisions.

As these countries grope toward political and economic arrangements that would answer to their traditions and needs, they are not likely to adopt a system exactly or nearly like those already in existence elsewhere. The conjunction of the ideals of freedom and justice leaves the form of the preferred political system quite open and does not in principle exclude some form of socialism. The only requirement is that people's basic freedoms are not encroached upon both in the choice of the political system and in the development of social institutions. The great insight of the founding fathers of the American democracy is that the allegiance to the state, whatever its particular form, must also be limited—it cannot bind the citizens' consciences and should leave intact their capacity for autonomous judgment and its expression.

The dialogue between East and West, Marxism and capitalism, has a chance of being fruitful if its vocabulary allows one to point up the connection between social ideals and the concrete ways in which the pursuit of those ideals affects human lives. If that connection is constantly, honestly, and vividly insisted upon, it is not unlikely that at least some rational, dogmatism-resisting adherents of Marxist ideology will not remain blind to the drawbacks of limiting freedoms. Having become aware of such drawbacks, they may begin looking around for ways in which the goals of justice and of freedom can be simultaneously pursued, as appears to be happening in contemporary China. For this highly desirable development to occur in the Soviet Union, its leaders must be courageous enough to take a step that country for so long has been afraid to take, namely, to open its frontiers to the flow of ideas and people. While such a step cannot be forced, it can come about as a result of a free debate among the Soviet leaders themselves; to be genuinely effective, it must be a decision made by persons in charge. Such a change, however, can be facilitated by a rational persuasion from outside, and therefore it is up to Western democracies and to more independent Marxist states to keep holding up, not in invective but in persistent argument, the mutual desirability of opening an exchange of views on matters so far shied away from. Before any other kind of disarmament can occur, there must be a *disarmament in ideological warfare.* Such a disarmament does not entail the abandonment of all of one's views and practices, and certainly is no threat to the maintenance of sovereignty. If force is foresworn, there will be no *coercion* to do anything; whatever actions are undertaken will come in the wake of being persuaded that they are in one's own interest.

A part of this persuasion process would include pointing out that the communist countries which allow ideas and people to circulate do not get thereby politically undermined. Yugoslavia has had an intensive contact with all countries all over the world, communist and capitalist, and still steers her own course. The same would happen in other Eastern European countries and most assuredly in the Soviet Union if it opened itself up to the rest of the world. The fears about subversion are the result of inertia, of failing to perceive that the policy of self-isolation is counter-productive. The benefits of breaking out of a rigid, inflexible posture include the prospects of establishing a peaceful and friendly relation to the rest of the world. In the light of these benefits, the risks turn out to be exaggerated or wholly imaginary, a remnant of a tradition that has outlived its usefulness and, in fact, is the source of untold harm.

If the prospect of benefits resulting from genuine trusting communication is persistently and tirelessly presented, if it is backed by rational argument, and accompanied by demonstrable and believable expressions of good will, it is not impossible that at least some influential persons among the Soviet leaders will respond to its appeal. A tentative interest in the idea can gradually grow into a constructive acceptance, enabling energetic and autonomous persons on both sides of the ideological divide to start creating paths toward mutual accommodation and cooperation, with freedom and justice for all as the ultimate goal. That spirit of cooperation is likely to spill over into other areas of conflict in the world now fed by intransigent nationalistic and religious allegiances. By establishing a genuine dialogue between themselves, the two ideologically opposed super-powers can set a contagious and persuasive example to the less massive particular antagonisms all over the world. By speaking out to one another on matters that affect the well-being of every person on both sides of the ideological divide, the mutual isolation can be broken and the prospect of attaining some degree of harmony may become a real possibility.

The Personal Factor

It is disheartening to me to see that ideological, nationalistic, or religious fervor is carried to the point of sacrificing personhood, one's own and that of putative enemies. It would seem that after millenia of warfare inspired by tribal allegiances human beings would see the fallacy of the needless destruction of precious human lives at the altar of parochial values. Today men are offered the possibility of contact with representatives of nations, religions, and cultures other than their own, enabling people to get to know, understand, and appreciate the validity of alien points of view. It therefore, seems surprising that it is still possible for people to be excessively and fanatically identified with their own group. In recent decades, this fanaticism has been especially virulent in political ideological allegiances. This new form of tribalism is more dangerous than the older ones because it is primarily intellectual in character—a *philosophy* of communism or fascism captivates the minds of their adherents, although such ideologies are frequently grafted on nationalistic values, as the nationalistic values are often combined with religious allegiances.

And yet, when members of opposing groups meet face to face as individuals and approach each other as full persons and not as mere spokespersons for their group's point of view, they can become genuine friends and even admirers of one another. It is due to such personal contacts that distortions and misconceptions about the members on the other side of the national, religious, or ideological fence are tellingly and convincingly removed. One comes to realize that beneath particularistic forms reside persons like oneself, capable of impartial objective judgment, morality, and autonomy. The essential equality of all members of the human race has been eloquently proclaimed by many philosophers and religious thinkers, ancient and modern; the brotherhood and sisterhood of all human beings is the cornerstone of Stoicism and Christianity and has provided impetus to revolutionary political movements in the doctrine of human rights equally possessed by all persons.

In the increasingly interdependent world which I now am, more and more institutions and organizations—economic, social, cultural—come into being as a result of the initiative of autonomously motivated individuals to pool their resources for the common good of its volunteer members. In such institutions, individuals refuse to delegate authority to impersonal, anonymous forces; they take responsibility for the way common projects are pursued. Such institutions can cross ethnic, religious, and ideological boundaries because the values securable by cooperation transcend and enhance those available from the most limited commitments. These kinds of social relations are more rational and more fulfilling precisely because they can help bring to realization the full resources of human beings. They are not constrained by parochial allegiances and hence liberate wider creative energies of individuals, thereby setting in motion similar creative responses in a greater number of people.

One telling example of this kind of occurrence is the ability of great literature, art, and music to appeal to persons inhabiting all corners of the world. Another example even more observable, is the internationalization of science and technology. Like other cultural goods, the fruits of scientific, medical, and technological discoveries are transforming the lives of more and more inhabitants of the global village. Nothing is more heart-warming than a sudden realization that the barriers dividing persons into hostile camps are not inevitable but accidental and artificial, mere vestiges of tribalism dressed up in faded costumes of parochial ethical, religious, or political allegiances. I am confident that persons who experience such an unveiling of the personhood of others up to now regarded as alien or pariahs will count this discovery as liberating and exhilarating. Such experiences reveal my ardent desire to exist at the level of personhood and to enjoy the potential blessedness of this level in as many human beings as possible.

As in all other instances of advance beyond the limitations of established standards, the origination of reconstruction and reform lies with individual persons. All significant changes depend on the initiative of individuals who, capable of non-routine reflection and judgment, pinpoint problems and propose solutions. Taking initiative requires not only independent, autonomous judgment but also courage. Among many persons who exposed the grievous shortcomings of the communist ideology, the name of Alexander Solzhenitsyn stands out. He dared to speak on behalf of the millions of human beings whose humanity was trampled upon by the atrocities of the Stalinist state. In his writings, he graphically depicted the magnitude of the oppression visited upon the inhabitants of the Gulag Archipelago, an enormous complex of camps maintained by the Soviet state to punish political dissidents and to isolate them from the rest of the population. Solzhenitsyn's condemnation of the system that made such inhumanity possible rang out eloquently in his literary and polemical writings, causing a serious concern among the keepers of the system. Afraid to face openly his charges against the system that it deliberately kept closed to any internal or external criticism, its keepers decided to exile him. Expelled from his native land, he continues to expose the cruelty, terror, and the moral corruption of a political regime that in the name of the supposed glorious future of humanity tramples on the lives of countless innocent human beings. Solzhenitsyn's stand is a telling denunciation of the totalitarian slogan that the end justifies the means. It effectively demonstrates that good ends cannot be attained if the means to them entail a colossal inhumanity of man to man.

Solzhenitsyn's life takes on substance from the cause to which he is dedicated. His stature as a person stems from the magnitude and the difficulty of the task he has chosen as his own. He is an impressive counter-example to the theory that individuals are molded by their society. By setting himself against the system in which he grew up, he forged his personal integrity out of his own judgment about that system. He found its practices sinister and destructive, and, merging his own suffering with that of millions of others, embarked upon a precarious and excruciatingly demanding course. In a lonely fight against overwhelming odds, he challenged those who arrogated to themselves the right and the power to impose immense suffering on multitudes of persons in the name of the supposedly glorious future. If one is inclined to say of him that he is larger than life, that characterization derives its plausibility from the fact that his life has become more meaningful because of his large task. His identity as a person is not separable from that task.

A similar degree of individual commitment and courage was manifested in Martin Luther King, Jr.'s struggle against racial discrimination in the United States. He saw his life in terms of his connection to Black Americans and he took upon himself the task of defending their dignity and human rights. He sought gains and changes in American society that at the time looked tremendous and overwhelming. Nevertheless, convinced of the importance of his task to end the disabling discrimination against millions of people, he did not shun effort or sacrifice, eventually paying for his cause with his life. By tying his own destiny to the natural longing and aspirations of persons who happen to be born with black skin, King made his personal identity much more weighty than it otherwise would have been. In remaining persistently loyal to his objective, he gave substance to his personhood. He chose to understand his destiny in terms of that self-imposed task. He had a dream: to lead Black Americans to the promised land and the mountain top. Although that dream is not yet fully realized, King's spectacular accomplishment in the area of civil rights and race relations is undisputed. A martyr to his cause, he has an assured place of honor in history.

Mother Theresa of Calcutta commands admiration because she has dedicated her life to alleviating the hardships of the poor. Struck by the ubiquitous poverty and misery around her, she concentrates her energies and abilities on doing all she can to lift as many persons as possible out of their enslavement to physical cares. By managing to reduce poverty and suffering in countless lives, she makes these lives better and happier. Her absolute commitment to that cause did not remain unnoticed; she is not only a celebrity but also a recipient of the Nobel Peace Prize. The honors bestowed upon her are a tacit acknowledgement that the work she is performing is deemed of utmost importance by those who are moved by considerations of love and justice. Her personal war on poverty makes sense because her success enables countless people to lift their eyes from the constant cares of the body and to exist at least intermittently at the level of viable personhood.

These three individuals are singled out because they spoke out vigorously and effectively against the disabling conditions of personhood. Their success proves that it is through efforts of *individuals* that changes in social and political life of humanity can be brought about. Through their lives, they give witness to my desire to exist in persons at the level of autonomy and blessedness, potentially available to *all* human beings.

4

My Greatest Danger:
Demise of Personhood

❧ The Shocking Thought of Extinction ❧

Weighty and morally important as the special disabling conditions of personhood are, they pale in comparison with a problem that has begun to penetrate human consciousness in recent years. A new and unprecedented worry has descended on humanity. Suddenly people realize that the wonderful world disclosed to them in their multi-faceted, absorbing experience may disappear. It may disappear for individual persons, for nations, for the human race, and even for other forms of life on earth.

The significance of the increasingly real threat of nuclear war is finally beginning to sink in. At stake is survival or... extinction. The break in meaning between these two nouns is awesome. I cannot even say that they belong to two different spheres of *meaning*, because the contrast between life and death is absolute. It is the difference between meaning and meaninglessness, sound and silence, reality and nothingness.

When I contemplate the no longer implausible thought that all or most of life—certainly the human kind of it—may disappear from the earth, I begin to see that individual and social existence makes sense not only because humanity has a past but also only if it has a future.

Human existence is suffused with the massiveness of its history. Persons feel connected to the stream of life through their families, communities, ethnic and cultural traditions, all ultimately embracing the entire career of humanity on earth. Beyond human history there is also the history of life on earth, with its absorbing drama of proliferating forms. Every little personal drama has its larger context. The projects and values that make human lives worth living seem real only because they are propelled by the desire to persist, change, grow, take a different, unforseen shape in the lives of children, of future generations. This sense of the continuing future, suffused by hopes, projections, and expectations, gives weight and substance to human existence *now*. The shock comes when this version of ongoing time richly packed with future possibilities is replaced by... nothingness. The very reality of the human way of being, of the accumulated meanings and values it exhibits, vanishes into thin air. Denuded of human life, I revert to being colorless, soundless, cheerless, meaningless.

⅍ The Message of Nuclear Silos ⅌

More and more people in the world begin to fear that their continuance as persons is directly threatened by the runaway proliferation of nuclear weapons, now amounting to 3½ tons of explosives for every person in the world. For a being whose very mode of existence involves an essential relation to the future, the contemplation of a real possibility that there may *be* no future is psychologically devastating. The sense of desolation, radical disorientation and abandonment has been tellingly captured by Jonathan Schell in his book *The Fate of the Earth*. The evidence collected in that book makes it clear that for the first time in history the possibility of the demise of humanity on earth is no longer incredible. It is not just the darkness of *personal* extinction, it is "a darkness in which no nation, no society, no ideology, no civilization will remain; in which never again will human beings appear on earth, and there will be no one to remember that they ever did."[13]

Those who attempt to picture some kind of survival after the nuclear holocaust overlook the fact that the achievement of an integrated, well-functioning personal and social life requires as its background certain minimum conditions of order and civilized existence. There is ample evidence, accumulated during the most recent wars, that under some conditions human beings lose the capacity to act as moral, responsible personal agents. Their very personhood is threatened. When physical deprivation and inhuman treatment reach the levels described by survivors of prisons and concentration camps, behavior based on civilized standards is made virtually impossible. Driven exclusively by the survival instinct, human beings not surprisingly lose both the interest and the ability to act in the light of such standards. The minimum conditions for the use of these standards do not exist when all respect for human life and dignity disappears from the scene. I marvel at heroic individuals who manage to preserve at least some of their humanity and personal integrity under conditions of extreme deprivation and torture. But I also recognize that morality, with all its cultural fruits, can be upheld only under mutually maintained conditions of solidarity and cooperation. A nuclear holocaust would make the maintenance of such conditions impossible.

It seems, then, that in order to preserve the viability of personhood, humanity must set itself with all possible determination and vigor against the mindless proliferation of nuclear weapons. The unprecedented character of what is at stake is described by Schell in chilling, mind-boggling statements: "Formerly, the future was simply given to us; now it must be achieved. We must become the agriculturalists of time. If we do not plant and cultivate the future years of human life, we will never reap them."[14] The present generation can look on itself as "a delegation that has been chosen by an assembly of all the dead and all the unborn to represent them in life." "...the people of the present generation, if they acquit their responsibility, would be the oldest of grandfathers, and their role would be that of founders."[15]

❧ My Future: In the Hands of Persons ❧

Considering that personhood emerged in a slow, prolonged experience of the human race, literally lasting millions of years, the responsibility for being the actual determiners of whether or not there will be future generations is awesome. It rightly fills many people with unprecedented anxiety, with cosmic bewilderment.

They must overcome that bewilderment, and set themselves soberly and whole-heartedly against the madness of multiplying the means of self-destruction. Repeatedly I have pointed out that the standards and ideals by which people live receive their content and function from the way they are understood. Their rational validity and viability depend on articulation and constructive interpretation by specially placed persons. These persons are often in a position to exert sufficient political and moral power to determine how seriously certain standards of behavior are regarded and how they are employed to affect policies and legislation. Among these policies are military options, including nuclear warfare, and these options are in the hands of particular individuals.

This being the case, to prevent the potential release of earth-destroying weapons in the world, people must mobilize all their resources to put moral and political pressure on persons who control the production and the potential employment of nuclear weapons. One task of this mobilization is to make clear to leaders of countries in possession of nuclear bombs that they do *not* have the mandate to invoke some supposedly independent and objective social or historical forces as allowing them to contemplate the option of using atomic weapons. I assure you that there are no such forces. There are only humanly articulated ideals and strategies to bring about certain results. The expression of such ideals and strategies must be rational enough to *exclude* options that threaten the destruction of the human race. Considering how much time, effort, ingenuity, sweat, blood, and tears it took for humanity to emerge, its task of saving itself is relatively simple. As Schell puts it: "Our modest role is not to create ourselves but only to preserve ourselves."

The individuals who are so placed as to be able to tilt the fateful decision one way or another on this question cannot remain anonymous. They must be identified, named by name, and reasoned with in an intense, insistent, and relentless way, in the hope that they will understand the consequences of these momentous options. In order for this moral pressure to be effective, everyone who realizes what is at stake must contribute to the generation of an intellectual and emotional climate in which the search for a way to bring about nuclear disarmament can go on. In that sense, the blame for the failure to generate a climate which would vividly demonstrate the folly of nuclear confrontation is to be placed on every person aware of the problem. By failing to persuade one's leaders to abandon the dangerous path, one contributes to mankind's potential suicide.

The task of alerting those in positions of leadership to this unprecedented danger calls for the exercise of imagination, ingenuity, initiative, and patience. They cannot get off the hook by merely pointing to obstacles and to barriers of communication. Citizens of one ideological or national camp must try to reach not only their own leaders but also those on the opposing side, ignoring or bypassing the deliberately or unwittingly maintained communication barriers. There is no other way to bring about a peaceful change. Since all policies and decisions ultimately are originated by persons, every effort must be made to reach those persons whose views, decisions, and policies count the most. It is not enough for the like-minded to talk to each other. One can count on the occupants of the White House to be generally like-minded, and similarly on the occupants of the Kremlin. But these self-imposed mental ghettos are irrational, and one must patiently try to penetrate the consciousness of people who, for whatever reasons, may be blind to the horrible consequences of policies they are prepared to pursue.

Such attempts to bring the leaders of opposing camps to their senses are more likely to succeed if issues other than those directly related to nuclear armaments are at least temporarily bypassed and if the negotiations concentrate fully on one paramount goal: termination of production and eventual dismantling of nuclear weapons. This goal should not be linked to other beliefs and values of the people on the opposing side of the ideological fence. Differences in the perception of what is politically, socially, and morally desirable will remain, and they can be dealt with in various ways, depending on the nature and complexity of the issue.

To solve the problems of the world, there must first *be* a world. Problems other than those having a direct bearing on survival allow for opposition, competition, contest, even a degree of mutual isolation. But they cannot be dealt with at all if the opposing parties destroy each other. Thus, even to assure each other of the possibility of remaining ideological enemies dictates the solemn agreement to refrain from blowing each other up by nuclear warfare. There is also a possibility, however, that, seeing themselves in this unprecedented and potentially catastrophic predicament, the opposing ideological camps will be more willing to soften their respective stands even on some fundamental political issues and will become more receptive to each other's views. The threat of mutual death may bring them closer together in life. "When it is dark enough, you can see the stars," observed the historian Charles A. Beard.

The biggest question ever confronted by me since discovering life is whether those who now run the affairs of humanity are sane enough to allow their desire for survival to inform their thinking and decisions. They have no other rational choice than to intensify their efforts to conclude disarmament agreements and to live up to them. At present these efforts seem pitifully half-hearted and minimal, given the grimness of the danger I face: the real possibility that life on earth will be destroyed. Never before has humanity faced a more momentous challenge. The horrifying thought is that it may be its last. It is horrifying to me because I see in life not just a sheer biological force but a glorious cosmic phenomenon that has produced the values of personhood. It is for this reason that I would like to mobilize all persons in what is essentially a struggle against self-destruction. The danger I face does not lie in inexorable and implacable forces outside human wills. The outcome depends on decisions of specially placed persons, and they must be shown that the fate of life on earth is in their hands.

I want to impress on human beings the fact that none of their decisions can be deferred to some nonhuman or superhuman agency or entity. People alone are responsible for dealing with the grim threat of nuclear extinction of life on earth. How should they live with the thought of this threat?

One possibility is that as this thought sinks in, it will lead to reactions typical of severe mental stress: depression, disorientation, panic. According to some psychologists, this response is already happening. Subconsciously convinced that there will be no future, people begin to see no sense in family life, in marriage, in raising children, in pursuing knowledge, careers, or art. A twenty-five year old woman contrasts her present mood with that of her childhood:

> The difference between those nights and these nights is that now there is no morning. The sun does not dispel my dread, and no one—not even my mother and father—will tell me that no killers are lurking in the closets or outside the window. The killers are, indeed, in the closets—closets called silos. And they are not only outside my window but surrounding me everywhere.

In a recent book, Lewis Thomas, a renowned scientist, confesses that he cannot put up with a thought grinding its way into his mind when he thinks of these closets called silos. "How do the young stand it? How can they keep their sanity? If I were young, sixteen or seventeen years old, I think I would begin, perhaps very slowly and imperceptibly, to go crazy."[16]

The sense of hopelessness is conducive to the abandonment of normal satisfactions in favor of artificial, perhaps drug-induced states of oblivious euphoria or of unbridled headlong pursuit of extraordinary thrills and pleasures. Eat, drink and be merry for tomorrow we shall die. When the student population of a major university votes in favor of stockpiling cyanide pills to be swallowed upon the approach of a nuclear holocaust, the reason may be a rational desire to influence the government policy, but that desire may verge on hidden despair.

One thing is clear. If people succumb to the mood of despair, they will hasten the holocaust by encouraging it in their own souls and in those already inclined toward reckless, thoughtless, irresponsible behavior. Affected by this mood, persons in charge of political and military affairs are more likely to make irrational fateful decisions. To some it may occur, however, that the actual amount of time available to humankind makes no difference to how one ought to live. Whether the time at the disposal of humanity is extended or short, it is a mistake to allow full sway to lowest, weakest, and meanest impulses. As long as people have life they will make that life more meaningful if they put to use their *best* capacities. Their highest satisfactions come from making use of creative and productive powers. History shows that the use of these powers enabled humanity to struggle effectively both with evil impulses in itself and with the hardships put on its path by indifferent nature. The result of the use of these powers is something I have never seen before: the flourishing of human culture with its various forms of knowledge, morality, and art. Any sense of pride or dignity I may claim for the human race is due to these achievements, and the real possibility of losing them forever underscores their irreplaceable worth.

Even in the darkening shadow of the mushroom cloud, persons can affirm the preciousness of human life by trying to live up to what they *know* to be the best in them. Should they fail to avoid the nuclear catastrophe and the concomitant extinction of the human race, they will at least die with dignity. Even in the face of imminent demise, it is good for me to know the worth of human existence in the midst of the otherwise meaningless silence of my lifeless cosmic spaces. This knowledge would be at least a tiny victory over death. Pascal's famous words come to mind here. "Man is but a reed, the most feeble thing in nature, but he is a thinking reed. The entire universe need not arm itself to crush him. A vapor, a drop of water, suffice to kill him. But if the universe were to crush him, man would still be more noble than that which killed him, because he knows that he dies and the advantage which the universe has over him; the universe knows nothing of this."

Pascal's last sentence is an instance of "pathetic fallacy." I never conspire against my most favorite mode of being, and, if it fails me, it will be due to the arrogance, recklessness, and stupidity of particular individuals. The physicists are right when they remind themselves that some day the lifegiving light of the sun will cease to support my experiment in living on the planet earth. The nuclear explosion set off by human beings would merely accelerate the process. But in either case, whether the course of life on earth is long or short, I have been enriched and made more interesting by lives of persons who managed to live up to the full scope of their personhood. My hope is that within that scope there may be a way to avert self-destruction. And then the human race can go on for a few million years. Who knows, perhaps it can even find ways to export personhood, the most advanced form of life, to other regions of my cosmic spaces.

My refusal to contemplate the demise of persons with equanimity comes from the realization that they are anxious to express themselves, and me, in solving some of the most fascinating problems of human history. Human beings all over the world now face many live and monumental options. Here are some of them:

1. In many thoughtful persons I feel a burning curiosity about how the present ideological conflict between East and West, communism and capitalism, is going to turn out, *if* it is not prematurely terminated by nuclear explosions. Will one or the other side win outright? What would be

the consequences of either victory for the future of humanity on earth? If socialism wins, is it bound to resort to repression to maintain itself or can it develop a "human face"? If capitalism wins, will it succeed in preventing abuses of economic power and in controlling its booms and busts? Or will the contest lead to a convergence of the two systems, synthesizing ideals and objectives championed by each?

2. How will the struggles in the Middle East, in Latin America, in Northern Ireland turn out in the end? Will the peoples in these regions of political and religious turmoil manage to resolve their differences or will they continue to tear apart their common humanity in bitter and senseless warfare?

3. How will the threat of overpopulation be dealt with? What solutions will be found to secure adequate supplies of food and fuel for all inhabitants of the earth? Will humanity put a stop to such tragedies as the current famine in Africa? Do the long-range possibilities include, as some enthusiasts predict, a colonization of space?

4. How serious can be the consequences of lapsing into decadence and defeatism which periodically accompany the decline of civilizations? Can the descent into immorality, reckless self-indulgence, and crime reach high enough proportions to destroy all personal autonomy, sense of responsibility, and creative aspirations?

5. What further changes will be introduced into human life and the planet's ecology by new scientific, medical, and technological breakthroughs, and by more effective and extensive methods of education? How may these developments affect the health, the capacities, modes of life, and well-being of people?

6. What consequences for the entire human race will result from a real in-depth contact among different cultures and civilizations of the globe? Is it not possible that the world views and religions—the Western, the Indian, the Islamic, the Chinese, and the African—which so far have been confined within their own geographical domains—will affect and fructify each other in totally unforeseen ways? Will perhaps a new global civilization emerge in the process?

These are but some of the questions that I find agitating the consciousness of many thoughtful and inquisitive persons. How they will be dealt with, with what degree of commitment, and with what resources and methods, is again up to persons, as individuals and in cooperative communities. With trust, but not without trepidation, I deliver my fate and my future into their hands.

5

My Religious Quest

⚭ Personhood in Religion ⚭

Ultimate Concern

Looking back at my adventure with the form of life, I cannot disguise a sense of satisfaction and pride. That adventure took me from blind experimentation with meaning displayed in the minimal transactions of simple organisms to the full-blown enjoyment of meaning open only to persons. The rudimentary type of value realized in the activities of an amoeba was magnified a thousandfold in the activities of human beings as they came to express the highly integrated, complex values of personhood: rational activity, integrity, autonomy, morality, mystical vision, and blessedness. Indeed, the quality of life exhibited by human beings is so different from and so much more interesting than the modes of being characterizing all other entities inhabiting my immense spatial expanse that the overwhelming sense of loss which accompanies the contemplation of a possible demise of personhood fills me with dread. The shocking thought of extinction does not reflect merely subjective feelings of life-loving persons; on the contrary, that thought takes realistically into account the objective catastrophe affecting the whole universe. I am aggrieved and anxious when I envisage the possibility of being reduced to prehuman levels of existence.

In this attempt at autobiography, I am trying to indicate just what is special about my discovery on one tiny planet and why the experiment fires my imagination. (Of course, it can fire it only because the experiment itself is responsible for the very possibility of imagination!) That the experiment is still going on and that it confronts many difficulties, problems, and uncertainties is no secret. The topics discussed earlier show that I—in the reflections of representative thinkers who catch the drift of my possibilities—am very much aware of obstacles still to be overcome. At the same time, however, the very presence of obstacles and problems testifies to the ongoing interest on my part in seeing the experiment through to its optimal opportunities. Just what these opportunities are is, and will remain, an open question as long as the drama of life is allowed to go on. The outcome of this drama, I have admitted, is largely in the hands of persons; their decisions and actions will determine, first of all, whether the human race (and much of other life) will survive and how it will continue to fare in my cosmic domain.

My present chief concern is the one just voiced: the danger of a nuclear holocaust. The thought is abhorrent because it connotes the loss of most important values, the values of personhood. The acknowledgement of the ultimacy of these can be seen in the phenomena of religion. It is time to bring them into the picture. Although religion has not been explicitly considered so far in this autobiography, it was present all along as its background. In telling my story, I was concerned to discriminate among various kinds of values displayed in my career. The result of that discrimination was the emergence into prominence of those that appear to be highest, most worthy of attention and propagation, that is, values that ought to be the focus of what Paul Tillich calls "ultimate concern."

Since I do regard some values as deserving this status, I could not be described as indifferent to the religious quest. In fact, as will soon become apparent, the central concepts and doctrines of religion are derived from values discovered in the phenomena of personhood. Although many religions were developed in the course of human history, they contain a common core which assigns special importance to activities characteristic of human beings. It will be easier to demonstrate that connection by concentrating on one major religion, so I shall draw primarily on the beliefs of Christianity, occasionally referring to other revealing manifestations of my religious quest.

The Roots of Religion

In the earliest forms of religious thinking, gods were imagined to be human beings writ large, each particular god controlling a specialized domain—the sky, the ocean, the rivers, the woods, and the fields. Divinities were unabashedly anthropomorphic. Since manifestations of nature are multiform, it is not surprising that early religions tended to be polytheistic. All events in nature, including movements of the sun, the winds, and the sea, all changes in weather, harvests, prosperity, illness, and disasters were explained by reference to direct action of particular gods. The relations among gods were also conceived in a human manner, depicting alliances, quarrels, and disputes over domains, prerogatives, and hierarchies.

When Plato reflected on the behavior of divinities in Homer's pantheon, he did not find the spectacle edifying. Plato's moral sensitivity led him to conclude that Homer was telling lies about gods and that only *one* Divine Artificer, or Demiurge, could be responsible for the order and beauty found in the world. Likewise, the experience of Hebrew tribes, fighting one another in the name of particular divinities, induced some morally outraged thinkers to opt for a monotheistic picture. The belief in one God made it possible to champion the underlying unity of the warring tribes and later, when developed by the Stoics and by the Christians, underscored the unity of all peoples inhabiting the world.

The path to monotheism was both a moral and a cognitive breakthrough. It raised the possibility of an overview of the entire universe by a being, God, who is concerned about the values to be realized in the universe. The monotheistic conception facilitated an intellectual unification of all subjects of possible knowledge with a concomitant affirmation of values realizable in the cosmic scheme. Since the events and processes of the universe appeared so vast, overpowering, and mysterious, it is not surprising that different cultures produced different views, some of which became great world religions.

Despite differences in their conceptions of God, all religions repeat a common refrain: to characterize God's essential reality, all of them draw on the idea of *person*. It should not be surprising. In fact, the idea of the Perfect Individual is a natural culmination of the original discovery of personal individuality. It is quite clear and noteworthy that such an individual would manifest four capacities attributed only to persons: 1) the ability to act, perform, create; 2) the capacity to care for and to love the things created; 3) the desire to revise, heal, and repair things that go awry; 4) the concern with the outcome of the activities pursued.

Central features of personal individuality are continuous with and rooted in my other life phenomena. 1) Even simple life has a capacity for spontaneous movement, for rudimentary action. 2) Every living entity cares about itself and about the surrounding that affects it. 3) The *conatus* of life expressed itself in adjustment and modification of behavior, in time giving rise to instinct—a species intelligence. 4) All forms of life seek situations which put them into a state of contentment. When extended to the level of persons, these features became self-conscious, in virtue of the linguistically acquired capacity for reference and self-reference. Hence, persons are 1) deliberate agents, 2) who know why, for the sake of what values, they act and organize their activity, 3) who are capable of taking corrective and self-corrective steps, and 4) who can offer a rational judgment, an evaluation of their pursuits and activities.

It is noteworthy that the Perfect Person, God of the Judeo-Christian religion, also manifests the four capacities attributable to persons: 1) as Creator, the Great Agent is quintessentially capable of producing acts of initiation, origination, spontaneous performance; 2) as the Loving Sustainer, He exhibits infinite capacity for caring, for unconditional love; 3) as the Redeemer, He intervenes to heal, revise and repair events and processes that go awry; and 4) as the Just and Merciful Judge, He is concerned about the outcome of projects undertaken and activities pursued.

A religious conception of life also attributed to God the Redeemer and Merciful Judge the capacity to bestow upon human beings a status that transcends their earthly career, thus giving rise to the idea of immortality. That belief once more testifies to the conviction that the value and meaning of human life is indeed ultimate in the scheme of things and, as such, is worthy of eternal blessedness. Immortality, then, is seen as the crown of personhood.

Creator: The Great Agent

A classical Greek myth reveals the need to ascribe God's actions to an intelligible motivation. Like human acts, divine actions must have a point. In his dialogue *Timaeus,* Plato claimed that God—the Demiurge or Artificer—had a reason for bringing the world into existence. God could not contain in himself his overabundant goodness and decided to share it with whatever was capable of partaking in it. In order for this plan to be attainable, He created an ordered world out of the primeval chaos of unorganized matter, looking to eternal Forms as a model. In attributing such a motivation to God, Plato made the right guess about my "state of mind" regarding the possibilities to be realized. He was correct in thinking that creative activity should make God feel good about Himself. The result of the act of creation was that the chaos of randomly operating matter was replaced by an intelligible, beautifully ordered world. God's satisfaction with his own goodness was enhanced when it was spread among the things and beings that could share in it to some extent.

In contrast to Plato, Aristotle did not think that the awareness of the world could contribute to God's contentment. Since the world contains imperfections, God's perfect repose could only be disturbed if He were to attend to them. Consequently, Aristotle described God's essential activity as the contemplation of His own perfection. God, the Unmoved Mover, does not love the world and remains wholly self-absorbed, while the cosmic process strives to imitate God's perfection. God's very presence inspires all things to strive toward sundry limited perfections of which they are capable. But the motivation is theirs, not God's. With much justification, some commentators on Aristotle have found his God aloof and self-centered. Nevertheless, God's very presence furnished motivation for all meaningful activity. Whatever is good *in* the world owes its goodness to God, and it does so to the extent that it approximates the kind of activity that is God-like in nature. Since the highest form of activity, according to Aristotle, is contemplation, thinking about thinking, this activity is to be sought by human beings as well. In doing so, they realize their essential nature. Analogously, beings and things can be called good to the extent that they realize their respective natures and live up to relevant standards. Aristotelianism, by emphasizing that any goodness is the function of accomplishing certain inherent ends, views my cosmic reality as valuable, because it displays activities which respond to certain standards of excellence. But this is precisely the capacity brought into existence by persons.

The Judeo-Christian monotheism sees God as the Perfect Individual, omnipotent and omniscient, guiding the world toward the goals He intended for it. He infallibly knows what He intends to bring about; there are no gaps or disharmonies between what He wills and what He does. To communicate His motives to humankind, God uses especially chosen human agents, prophets and messiahs, who reveal to ordinary mortals His will. The object is to bring the will of human agents into harmony with the will of the Great Agent—"Thy will be done." When properly informed through the commandments of the scriptures, human actions will enact standards acceptable in God's eyes.

When persons perceive their standards of action as a response to the action of the Great Agent, they may see a connection between themselves and God. The essential formal similarity is this: God is a being who can act in accordance with deliberately and autonomously chosen standards and whose mode of behavior is animated by His appreciation of the meaning and value of these standards. This is one sense of the expression "Man is made in God's image." Being made in God's image enables man to discern God's presence in the world and to become His collaborator toward achieving values He wishes to be realized. In such a collaboration, by proper thoughts and deeds, the sense of oneness with God is a possibility, a view which is a consequence of the insight that the divine and the human persons share similar or analogous standards.

The Loving Sustainer

A logical feature of a standard is that it cannot be enacted or followed only *once*. In virtue of being a standard, it lays down a precedent, a rule to be followed, and a rule that can be obeyed only once isn't a rule. Standards call for *sustained* activity. Once something is recognized to be good, worth caring about, then it is worth caring about on all occasions as well. One act of love calls for further acts of love. Thus love is a self-generating phenomenon; it is caring about caring.

There is an ambiguity in the notion of agency, human or divine. The agency may be episodic or sustained, and the ambiguity can be cleared up by distinguishing between *doing* and *being*. When a person performs an action in accordance with a standard, the doing of the action is an event, while the adherence to a standard says something about that person's being or character. Actions ascribed to God are seen as stemming from the goodness of His nature. That goodness, in Christian theology, is acknowledged in the doctrine of *continuous* creation, attributed to such thinkers as St. Augustine, Descartes, and George Berkeley. According to that doctrine, the world is not the result of just a single act but is constantly maintained in existence by God's loving will.

These two logically distinct types of acting are also ascribable to persons. Besides performing discrete actions that terminate immediately after being initiated, persons are also capable of *sustaining* actions. They are present in long range commitments, in principled behavior, in a "long obedience to a cause," to use Nietzsche's phrase. This mode of acting is especially visible in interpersonal institutional contexts — loyalty to persons and projects. As divine love or grace characterizes God's continuous influence, so a person's moral character or virtue characterizes a person's way of manifesting an active presence in his surroundings. Descriptions of character do not refer to individual actions but to a common element in them, as manifesting a steady attitude, habit, consistency, perseverance. In other words, there is a coalescence of doing and being; a person *is* what he or she habitually or usually *does*.

Like divine action through love or grace, human action stemming from character or stable disposition is *available* to its potential beneficiaries. It is not intermittently or sporadically or capriciously turned on and off. Rather, it can be likened to a reservoir of power to be drawn on when a need for it arises. A person may be available as a source of information, advice, material or moral support, as a teacher or guide. Like a human person, the Christian God responds to a request: "Seek ye and ye shall find." In a religious framework, as in interpersonal relations, a simple question or a request may be sufficient to activate a caring response. The response may be episodic or continuous, but, in either case, a personal intervention exhibits the feature which Socrates looked for when he claimed that the world governed by Mind is intelligible in terms of values and not in terms of merely physical, mechanical events. Mechanism explains, but value *justifies* by showing why something is good, worth bringing about.

When the agency, divine or human, is presented in other than personal terms, something less than justification is settled, namely, explanation in terms of impersonal causes. But God's personhood is hereby diminished. It is not surprising that deism became popular when modern science came upon the scene; it tended to explain all of my operations as automatic and thought of God as a celestial mechanic, a Divine Watchmaker who, having made the world like a watch, left it ticking by itself.

Similarly, when philosophers conceive of the divine cause as a metaphysical principle, such as Substance, Nature, or the Absolute, they make religion problematic. This happens also when a personal God is replaced by Spirit. By referring to God as Logos, Word, or Spirit, the Gospel of John assimilated the Scripture to Greek philosophical speculation, thus bringing some uneasiness into Christ-centered theology. Most religions feel more comfortable with the idea of a personal God because personal agency, by invoking the good which this agency brings into being, stops the potentially infinite regress of further explanations of that good and provides a conclusive justification for it in terms of its intrinsic value. Reference to personal agency has a kind of ultimacy that no other explanatory principle can supply. That is why it is found in accounts of human *and* divine behavior.

Joseph Campbell, an astute commentator on religious mythologies, has called attention to three different ways in which world religions represent the relation of persons to ultimate cosmic reality. The Eastern view sees the individual as phenomenologically different, and yet, in truth, *identical* with ultimate reality. The Western view emphasizes the role of a *mediator*—the Mosaic covenant of a people in Judaism, Jesus in Christianity, Mohamet in Islam. But Campbell also notes that the Europe of the Middle Ages has originated still another model, namely, that of approaching the mysterious core of reality through intense personal love, celebrated in the Troubadour and Minnesinger traditions. Personal love has the power to reveal a deeper dimension, an immanent divinity in nature. This kind of love Campbell calls *amor* and distinguishes it both from *eros,* an indiscriminate biological urge, and from *agape*, which is indiscriminate in another way—it sees *through* the individuality of the loved person to something beyond it. In contrast to both, *amor* respects the individual not as a member of some sanctified consensus but as having value in herself or himself.[17] According to this view, it is not universal features but the *particularity* of the person that is revelatory of meaning and value contained in the cosmos. It is instructive that in this conception of love the intensity of the relationship is not episodic but generates a continuous loyalty, a devotion unto death, as depicted in the celebrated story of Heloise and Abelard. In this kind of love, doing and being coalesce.

The Redeemer

In addition to being pictured as a loving, caring Creator, God is also thought of as a Redeemer. Creation and redemption are logically interconnected; in the end, they are concerned with the same things: to ensure that the outcome of the divine enterprise in the world is successful. First the creation is endowed with positive capacities, and then, should these capacities be destroyed or undermined, as in the doctrine of the Fall, the successful outcome can still be assured by the act of redemption—sinners can be saved. Even such an ultimate limitation as the inevitable human finitude is remedied through the granting of immortality.

In Christianity, the proclamation of man as made in God's image exists side by side with the picture of man as a sinner, a fallen creature. The fall from grace is attributed to the original transgression of Adam and Eve. Their story is presented as disclosing a universal truth about humans, namely, their inherent and ineradicable tendency to do evil. St. Paul lamented that no matter how good human intentions are, people often go willfully against them. A part of the religious message is to remind people of their tendency to sin, in order to induce repentance.

The notion of sin says something important about the way in which persons may view themselves. It points up the seriousness with which they can experience their shortcomings and transgressions. The concern with imperfection, failure, and evil is natural for a being who is concerned with making the good prevail. Therefore, it is not surprising that God, the Perfect Person, is seen not only as Creator but also as Redeemer, or Savior. If the divine creative enterprise is to be successful, there must be ways of overcoming sin and evil; hence, Christianity condemns transgressions against moral commandments, and exhorts believers to face up courageously and creatively to their evil tendencies. It also ascribes highest spiritual status to actions and lives that awaken the moral conscience by self-sacrifice and martyrdom, of which the story of Golgotha and of the Cross is the crowning example.

If creation is a primordial manifestation of caring, then love can be described as caring about caring. Redemption, in turn, may be characterized as a double creation or re-creation, a sustained concern to repair errors, mistakes, transgressions, failures. By conceiving of God also as a redeemer, Christianity provides a way of dealing with *negative* facts of the cosmos. It encourages certain appropriate responses to those facts. It presents God as compassionate, sorrowing over human transgressions, but also giving strength to overcome them. The Sermon on the Mount, for instance, tells the faithful what should be their attitudes to suffering and evil. They are not to be lukewarm toward these negative facts but must face them resolutely and steadfastly. When the evil is thoughtless, unreflective and brutal, it must be opposed by force, without sentimental illusions. When "turn the other cheek" has no effect, a Christian may be reminded that Jesus also brought a sword.

Nevertheless, it should not be forgotten that my discovery of life late in my cosmic career was the discovery of the possibility of *goodness*. Consequently, redemption from sin and evil is called for because they are the chief *enemies* of goodness. The faithful are involved in a fight against forces that *deny* goodness and virtue. Without the consciousness of positive objectives, the obstructions to them could not be perceived as obstructions. Therefore, in order to perceive in what the particular sins consist, one must be aware of the nature of the positive objective which the sins negate. Sins and transgressions are to be struggled against because they undermine human efforts to be guided by particular standards of goodness. Wickedness prevents persons from being able to achieve goals which, when attained, make their lives worth affirming and celebrating; thus it is proper to conclude that the notions of sin and evil, although certainly not unimportant, are nevertheless derivative. They are important only because the objective of life is to bring into existence various kinds of goodness and to celebrate its many manifestations. It is instructive to recall in this connection that, as the Latin root of the word "salvation" indicates, to be saved is to be restored to health, to a state of affairs which the Perfect Person would like to see realized.

Divine Judgment

Recall that the transition from instinctive animal intelligence to human rationality occurred when individuals became responsible for learning and enacting standards by which a human group could regulate its common life.

This transition calls for an independent judgment and scrutiny of the use or misuse or abuse of standards. Only autonomous persons can perform this task. The typical characteristic of personhood is the ability to look at things, activities, events, and processes in terms of their worth—do they manifest a value of a right kind? Upon reaching maturity, people can feel that it is their responsibility to subject to a critical appraisal values which come their way as candidates for enactment. The very possibility of taking or not taking this responsibility is an option that no less developed entity has, and that's why, in regard to life's future, persons are authentic and autonomous spokespersons for me. They are equipped to judge what I, as the original source, champion and upholder of life, am worth.

It is not surprising, therefore, that the possibility of judgment is also prominent in the religious idea of a Perfect Person or God. According to Christianity, as well as many other religions, upon completing their life human beings stand before God to receive a final judgment, an appraisal of whether they have acquitted themselves in the proper way. This notion of a divine verdict is also an application of a standard concerning the value of something, only in this case the entity to be judged is a life as a whole.

That notion is but an extension of what persons do all along: without necessarily being "judgmental" in the negative sense of the word, they pass judgment on their actions, designs, plans, and performances. Since this is a normal and natural attitude for a person to take toward his or her possibilities, it should not be surprising that the same concern with evaluation crops up in religious analyses of the human situation. It makes sense to ask what a life as a whole amounts to because persons are concerned with integrity and autonomy, with preserving a continuous sense of self-control over the course of life. If that life realizes positive possibilities open to it, the verdict can be positive.

One might say that such overall judgment, whether divine or human, is a second-order judgment, performed over and above, so to speak, on top of appraisals and verdicts passed on particular actions and events, seasons, stages, and stretches of life as they actually take place. The concern about the quality of one's life can engage a person as his life proceeds on its course, regardless of whether he does or does not hold religious beliefs. Religion merely intensifies the natural bent toward self-appraisal by invoking an allegedly authoritative court of judgment. In either case, persons show concern about their life's meaning. Although this interest may be temporarily evaded, it keeps reasserting itself, even if only intermittently, reluctantly, or half-heartedly. It cannot be suppressed altogether because to be a person is to be interested in what one's life amounts to in the long run.

The possibility of evaluative judgment, discovered by me in human beings, really rounds out my initial adventure of discovering life. In the initial and early forms of life, I had no capacities and no instrumentalities, such as the human brain, to understand the value and meaning of life. Since such understanding is the result of human appraisal, the authority of that appraisal may be put in question. The notion of divine judgment, prominent in most world religions, intrigues the human race because it purports to settle the question of authority. It lies in the very nature of value that it is curious about its own worth. That's why the existentialist slogan that human existence puts itself into question is only partially true; it leaves out the search for *answers* to this questioning. Such answers are provided by acts of valuational judgment, which, when positive, has its natural upshot and completion in the states of blessedness and celebration.

This phenomenon of human life is connected with the very idea of meaning and can be found at many levels. It is present already at the level of saying something, making an assertion. As I have noted, the use of language presupposes the respect for the standards of correctness and appropriateness of utterance. R.G. Collingwood suggested that every statement could be seen as a response to a presupposed question. Saying something must have a point; some need, even if marginal, must be filled. Apart from such need or point, a statement is literally pointless. In that sense, a statement, if it fills a need or makes a point, achieves a certain value, it rounds out, positively completes and resolves a "problematic situation." Many philosophers have been struck by the "dialectical" character of discourse, which typically consists of question and answer, saying and responding, putting forward a thesis, an antithesis, and then resolving their conflict in a synthesis.

Judgment as a rounding out, a completion of a context, can also be encountered as an integral part of a story. The tragic drama of King Lear had its beginning in Cordelia's refusal to complete the regal ceremonial of her father's bequest when she declined to offer a conventional "thank you" as he was about to give her one-third of his kingdom. Lear was incensed with Cordelia because, in contrast to her sisters, she failed to round out such a situation, to complete its value by saying "Thank you, Father." In that play, Shakespeare is demonstrating the importance of conventional completions for any civilized order. The value to which Lear was committed and expected his daughters to uphold was the value of gratitude. His gift was not complete until it was accepted as such. An objective sign of such an acceptance would be a little formal speech, an innocuous conventional game, which Cordelia refused to play, for reasons which were of interest to Shakespeare as a commentator on the complexities of human life.

The interest in producing "completing" judgments is related to the interest in integrity, autonomy, and blessedness. In the light of this connection, I am inclined to say that *the very idea of a person is the idea of a being capable of bringing values to their completions by acts of appraisal.* If the notion of appraisal is taken widely enough, persons can be seen as beings who are bothered by incompletions and by unintegrated, dangling remainders. It is not surprising, then, that the need to round matters out, to complete them by a proper evaluation, has found a prominent religious expression in the idea of divine judgment. That judgment puts a period to a life—by pronouncing on its worth. The deep urge to see things completed, to recognize them for what they really are, leads to the idea of a being, God, who can do it with authority and finality. The same need of completion prevented Socrates from being satisfied with the mechanical Mind of Anaxagoras because, as an explanatory principle, it was not capable of judging what is the good of that which is happening in the world.

A belief in Mind or God satisfies the urge to see the ways of the world justified and vindicated in terms of ultimate value. The conception of God as a Redeemer is invoked in order to ensure that no incompletions, no remainders, no dangling threads mar the total scheme. In A.N. Whitehead's interesting theology, God has both an antecedent and a consequent nature, the latter being precisely concerned with tying up loose ends and perfecting, bringing to completion, values that somehow failed to be realized fully. Aristotle's God performs a similar function—in him there are no unrealized potentialities. No further event or action is needed to make Him perfect. Therefore, He is outside time, immortal.

Immortality: The Crown of Personhood

Self-appraisal is most prominently represented in the work of moral conscience. Immanuel Kant regarded the voice of morality as the disclosure of the deepest resources of human personality. For Kant, the moral self is the *real* self. In his ethical theory, he even claimed to establish a connection between morality and religion. He argued that a person who acts morally deserves eternal recognition; moral uprightness is a direct path to immortality. The basis of this claim is that moral action is unconditionally good and, as such, merits divine reward. Indeed, Kant believed that from this feature of morality one can argue to the existence of God. The argument is relatively simple. Since moral action deserves and requires appropriate recognition, and since such a recognition in many cases is not given during a person's lifetime, one must postulate that person's existence beyond the grave. But in order for such an existence to be possible and for the deserved reward to be bestowed, there must be a being equal to the task. No other being is equal to such a task except God. Hence, the fact of moral experience proves the reality of God.

This famous "moral proof" of God's existence interests me because its intent is to reassure human beings that what they do in life matters absolutely. Kant's religious interpretation of morality places the uniqueness of human beings in their ability to act in accordance with and for the sake of the highest possible standard. That standard commands unconditional allegiance. The moral law, according to Kant, is the *supreme* good, admirable in itself, and its objective is to produce the *highest* good *(summum bonum)*, which happens when the actions performed out of respect for morality are also appropriately recognized. Immortality and God are postulated to assure the bestowal of this highest good, due to persons as the proper recognition of the moral intent of their actions.

What is important in Kant's analysis is that the moral action itself guarantees ultimate happiness, as he came to emphasize toward the end of his life. Virtue is literally its own reward. To understand the meaning and import of actions done out of respect for what is admirable in itself is to be in a state of mind that can judge correctly its own worth. Since that state of mind produces perfect contentment and peace, it could be called, perhaps more fittingly, blessedness. It accompanies moral action, even though other feelings may be present as well. The consciousness of having done what duty dictates can conquer the pain of thwarting immediate natural desires and inclinations. A person can know that even though the worth of good actions may remain unrecognized by the world, they *deserve* proper acknowledgement. A mark of self-respect is that it is not conditioned by whether or not one gets the proper applause. As Kant rightly observed, everything has a price, but only the respect for what is intrinsically and unconditionally good can bestow dignity. The sense of dignity, the conviction of its importance, is a prominent component of blessedness.

Another thinker who used the religious framework to emphasize the importance of ongoing self-appraisal was Goethe. At the end of his famous drama, *Faust*, the hero is saved because, as the Lord says in the Prologue in Heaven, "We can redeem anyone who keeps striving and struggling." Faust's mode of life receives divine approval; he has God's blessing from the very start. Knowing that Faust will act as God expects him to act, the Lord called him a good man even before the action of the play started.

Why does Faust deserve to be called "a good man?" His soul has a tremendous hunger for experience. But by "experience" Goethe does not mean a passive sort of awareness. On the contrary, he depicts an intensely active stance toward life, dominated by curiosity and creativity, openness toward the unknown and the strange, daring in the face of danger, fascinated with beauty and splendor. Such an attitude toward life includes the willingness to learn from mistakes, not to repeat the same mistake twice, and to suffer for one's errors and transgressions. What he expects from life, Faust describes as follows:

Whatever is allotted to the whole of mankind, that I wish to experience in
my inner self, grasp with my mind the highest and deepest there is, heap
upon my bosom all man's weal and woe, and thus extend my own self
to be one with mankind, and, like it, in the end meet shipwreck and
perish.[18]

The greatest enemy of experience conceived in this way is stagnation; it kills
the human spirit. The chief task of the devil in Goethe's drama is to do just
that, to kill Faust's multidimensional activism. The devil fails because
Faust's striving never ceases.

Faust is a good man in God's eyes because he takes seriously the divine
charge to make his life amount to something, and he considers the earthly
setting of his existence as the proper area for his activism. This setting is the
concrete condition for his spiritual striving. Faust's favored status in God's
eyes is due to his making a resolute use of the material conditions of his life.
In this sense, his human calling, which Faust perceives clearly only at the
end of his life, was obscurely known to him all along. As the Lord says, "a
good man, for all the obscurity of his impulses, is well aware of the one
right way." By responding to life's challenges and opportunities, Faust
expressed toward them an attitude which deserves to be called piety—giving
things their due and responding to them with serious interest and energetic
effort. To make his philosophical point dramatically, Goethe endowed his
protagonist with abilities that are larger than life; Faust is an extraordinary,
titanic figure. But the message of Goethe's drama is applicable on less
spectacular levels as well. For *every* person, the area of life is a stage for
spiritual quest. Building on Goethe's insight, one may say that wisdom
begins with the realization that the inexhaustible panorama of the cosmos
calls for an evaluative response. Only persons are capable of it in a full sense
of the word, which includes a serious attempt to arrive at an objective
judgment about the worth of one's activities in life.

At the height of their insight, both Kant's moral man and Goethe's Faust
know that they are blessed. They are conscious of themselves as facing
meaningful, ultimately valuable tasks, the fulfillment of which deserves and
receives recognition by the highest power there is. Goethe and Kant believed
that human beings who know their destination in life can be judged to be
worthy of immortality and salvation because they appreciate the full weight
of the challenge and the opportunity life presents them with. Their
knowledge is the condition of their correct appraisal of themselves and their
relation to the world. Their state of mind at the time of possessing that
knowledge can be called blessed, because they see their situation as
presenting objective challenges which call for a personal response. Their
situation parallels that of a religious believer who finds himself favored by
God's blessing and therefore sees his life as a gift to which he is enabled to

respond joyfully, with enthusiasm, and without reservations.

Both in Kant's theory and in Goethe's poem there is a strong concern with *accountability*. They recognize the importance of passing a verdict on the whole of a person's life. Does it *deserve* immortality? — in the case of Kant. Has it lived up to God's expectations? — in the case of Goethe. What is at work in both cases is the traditional notion of divine judgment. Interestingly enough, that judgment turns out to be a judgment of judgment. In Goethe's poem, God is interested in knowing how Faust understood and evaluated his own decisions and actions, how he judged the worth of his activities. God's verdict on the goodness of Faust's life as a whole includes God's taking into account whether Faust himself did value and appreciate what he was doing *as* he was doing it. God is pleased with Faust because he did not fail to keep an eye on the meaning of his experiences, reacted intensely to them and to their consequences, and kept track of how they cumulatively affected his soul, his whole person as an autonomous enduring entity. In other words, Faust did not live blindly but had his wits about him. That is why he was saved. Actually, his "salvation" was no more than God's approval of his way of living. He lived as a good man should, and God knew already in the Prologue in Heaven that Faust was a good man; i.e., would act in the way God expected of him. Faust's "salvation" is not an additional event tacked on at the end of his earthly career, rather it is an affirmation, an approving judgment of the totality of his life.

❧ Religion: My Cosmic Evaluative Quest ❧

The Common Core of Religion

Reflecting on the central tenets of Christianity lets me realize that they indeed translate the values of personhood into the language of religion. But this fact is perfectly understandable, since the religious refrain expresses my ardent desire to give the phenomena of value the greatest possible scope. The undisputed parallelism between some key religious concepts and the values of personhood indicates that the concept of God is but my mask for propagating these values. That parallelism is most prominent in the Judeo-Christian religion, but it is present in other religions as well. Even the Eastern religions which seem to emphasize the impersonality of ultimate reality cannot dispense altogether with the idea of person. Note, for instance, that the *Upanishads* begin with this statement: "In the beginning this was Self alone, in the shape of a person." Even more importantly, the claim that the supposedly impersonal ultimate reality is *spiritual* clashes with its characterization as impersonal. That spirituality without any personal characteristics is a most problematic notion is borne out by the fact that in Eastern religions the *representatives* of it are always persons, either the multiple and often warring divinities of the Hindu pantheon or its human revealers.

Another denial that ultimate reality is impersonal is contained in the claim of identity or non-duality of everything. The Hindu tradition dramatically amalgamates the personally human and the impersonally divine reality when it encourages the believer to say to everything "Thou art That." The locution is supposed to act as a reminder that God permeates everything. But if so, He is in persons and they are in Him. One can say "Thou art That" not only to a tree, to a river, but also to another human being.

Even more intriguingly, one can correctly say to oneself "I am That," where "That" stands for the value dimension of all reality. I cannot conceal a considerable attraction to this line of thought. By equating divinity with the entire scheme of things, that is, with me, the Hindu believer is at the same time calling attention to my value dimension. When he reminds himself "I am That," he does not say this self-aggrandizingly. On the contrary, the expression is a manifestation of piety; one is to realize whatever goodness one is capable of bringing about in virtue of one's connection with the cosmic order conceived as divine. This injunction is the message behind the seemingly extravagant saying "I am God." The point of the message becomes plausible if it is meant to call attention to the person's own place and time as encompassed by God's plans. To see the world and one's life in it as somehow touched by divinity is to feel called upon to bring about the realization of the good envisaged by God. To the extent that a person is capable of responding to that vision, it is up to that person to be God's agent. A failure to act from this injunction at all moments of life is not peripheral, on the level of some other errors. It is incomparable because, by definition, whatever is conceived as God-given is highest on the scale of values and should be a target of ultimate concern.

Adherents of Western and Eastern religions see their lives in the light of the conviction that human existence is in some ways touched by the divine possibilities of the world. This conviction inevitably leads to a thought that the points at which divinity touches a person's life are located in events and actions which that person judges to be valuable, admirable in themselves. In other words, to hear God's voice is to hear it addressed directly to oneself. That voice urges one to be a co-worker in the divine scheme, a collaborator in the project to realize highest possible goals, or at least to contribute to their attainment.

Religious concepts naturally grow out of the idea of human personhood.

The reason for the inclination to extend personal values in religious directions is not far to seek: I take these values seriously. Religion arises from assigning supreme importance to some aspects of personhood, and it consists in endlessly embroidering upon the awareness of that importance mythically, intellectually, imaginatively, poetically. Religiousness communicates the sense of wonder and even amazement discovered in deeper experiences and gives expression to ultimate human hopes, fears, and aspirations. The seriousness and openness with which persons treat the reality and the promise of life shows forth the prodigious workings of moral conscience and the prolific fertility of poetic imagination. That is why the religious quest has taken on such a variety of expression—from early primitive versions to the elaborate edifices of great world religions.

Because this probing into unfathomed aspects of personhood is so imperious and yet so elusive, a temptation constantly arises to capture it in some decisive doctrine and to reach a definitive resting point in some super-concept. The idea of God is such a super-concept, allegedly connecting human values with a supernatural reality. Every formulated religious orthodoxy is an attempt to put an end to the continuous, open-ended quest for religiousness. But the ongoing process of probing the potential depths or heights which life is exploring in humanity cannot be stopped by any super-concept. As long as life exists at the level of human personhood, it will continue the creative surge of intellect, emotion, and imagination toward greater self-comprehension and richer self-expression. To the extent that a religious system tries to contain or arrest this quest, it *limits* rather than expands my creative quest. Even the concept of God needs to be transcended when it becomes a stumbling block to that surge.

Paul Tillich shocked many Christians when he suggested, paradoxically, that one might have to reject religion *in the name of* religion. Similarly, John Dewey caused consternation in religious circles when he proposed that the noun "religion" be dispensed with, keeping only the adjective "religious." One need not be as radical as Dewey suggests; the noun "religiousness," properly qualified, is not an empty word. Tillich was on the right track when he proposed that religiousness be defined as ultimate concern. So was A.N. Whitehead when he claimed that religion is what persons do with their solitariness. Tillich's and Whitehead's slogans, when combined, converge on the notion of personhood. When the values achievable in every person's life are treated with ultimate concern, they are acknowledged as constituting an indefinitely perfectible challenge. To the extent that people can see me as a meaningful whole within which they can achieve autonomy, morality, piety, and mystical absorption, they can attain blessedness for themselves and for me.

Persons who take life seriously deserve to be called religious. They know that my quest for value is open-ended and never fully accomplished. Because the notion of *completed* perfection is inimical to endless quest, religiousness demands that quest be endless. It is pointless to impose arbitrary limits on the active, creative cosmic impulse which lies at the heart of that complex form of life manifested in personhood.

The Secular-Sacred Continuum

Because the values of personhood and their religious extension form an intelligible continuum, the possibility of shifting back and forth between secular and religious vocabulary is not to be frowned upon. Indeed, the very distinction between the secular and the sacred becomes dubious when the actual phenomena of life, those achieved in spiritual activities of living persons, are acknowledged as intrinsically and ultimately valuable. A person can *be* religious only in time, that is, on the secular plane of existence. To separate the sacred from the secular is to make the actual passage of time and all events in it devoid of spiritual value.

Religious feelings are a natural result of thinking seriously about deeper human resources. These resources may be understood and articulated more inclusively if no chasm is created between the secular and the sacred. The many parallels between the values of personhood and those acknowledged and celebrated in traditional religions testify to the *unitary* character of life. To think of it in this way is to find many forms of religious thought and practice congenial and enriching. One dividend of such an ecumenical attitude will be the possibility of drawing together persons who accept a particular set of religious beliefs in toto and those who limit themselves to acknowledging the personal human values which that set also celebrates. In both attitudes there is a convergence of what is undisputably worth championing, upholding, and protecting. Respecting each other's intellectual consciences, such persons will refrain from indulging in invidious distinctions and from labeling one another as "superstitious" or "scoffing." On the contrary, a person who accepts only the "humanistic" core of a religion will find in religious institutions and practices much that he will approve of and cherish on other grounds, moral or esthetic, for instance.

Even on a purely intellectual, philosophical level, a sympathetic attempt to understand one another leaves plenty of room for a constructively critical dialogue. A discovery that religious concepts have their roots in the very nature of personhood may help the tree to religiousness to blossom in future new forms. It is good to learn that religious feelings and attitudes do not descend unaccountably from a mysterious realm. Religion does not make people serious about life, but rather the relationship is the other way around: because people take life seriously, they are moved to give that seriousness multiple religious forms. My religious quest is a natural outgrowth of pushing the values of personhood toward their utmost limits. At all times, my nature tries to become *super*natural. My supernatural aspect, however, transcends my natural reality in the same innocent sense that an ideal always transcends actuality. In that same innocuous stance, the sacred transcends the secular. My perpetual miracle lies in the fact that through persons the mere appeal of ideals and of attractive new possibilities can galvanize me into action.

Religious concepts reveal ideal human longings for ever greater autonomy, happiness, and blessedness. They indicate the directions in which people must travel if they respond to the call of perfection that keeps haunting them. The idea of God is the idea of the resting point at which this quest for perfection would find completion. The ideal of progress, theologically conceived, is intriguingly similar to the ideal of personal growth toward maturity and wisdom. God is seen as the being in whom this growth is already perfectly completed.

Responsiveness to ideals, dramatically manifest in religious beliefs, has a less mystifying expression in people's responsiveness to extraordinary individuals—saints, heroes, and geniuses. Fascination with such figures betrays persons' awareness of themselves as value-sensitized beings, as entities who act in the light of norms and standards. People respect, honor, admire, and sometimes envy those representatives of humanity who have manifested unusually high degrees of personal mastery over themselves and over their circumstances. But they pay attention to them because they carry in themselves at least germs of capacities and abilities which especially gifted persons so gloriously display in their actions. Not to have an inkling of such abilities would constitute being deprived of a yardstick that enables people to judge the value of what they themselves can achieve. This is one sense of the ancient dictum: nothing human is alien to a human being. All people are members of a race whose special characteristic is the ability to conceive of and to act in accordance with norms that aim at a production of valuable experiences. Value, invented in early living organisms, has reached an intense flowering in persons who can consciously appraise and appreciate *what* is valuable.

Practical Consequences of Cosmic Religiousness

In this final part of my autobiography, I have reflected on the curious fact that religion has been a persistent component in the development of human cultures and civilizations. I have also noted that the religious quest is dominated by a concern for the values that have come into my existence with the advent of personhood. Religions insist that personal agency understands itself as being charged with the realization of highest possible values. But if this is so, one important practical conclusion can be drawn: every moment of a person's life can be seen as providing an opportunity for such a realization. When the judgments passed on one's doings meet the standards of what has worth, the result is blessedness. Since that special value is brought into me by persons, not to celebrate that value when it is manifest is an immeasurable loss.

Persons are my favorite mode of being because they are life's most individuated expressions. Each person's life is uniquely good in itself and makes a particular contribution to my quest for all types of goodness. An image may help here. If one compares the sweep of human experience to a meadow, then the good of each person is comparable to a flower protruding from a green grassy expanse. As the beauty of individual flowers helps to make the meadow lovely, so the individual fulfillment of persons makes me, the universe, beautiful. As the loveliness of the meadow is the function of the beauty of its flowers, so each increase in the well-being of individual persons increases *my* value. Persons are active contributors to the sum total of values of which I am capable.

I would therefore remind every human being "the action is where you are." If it is possible for you at any given moment to bring about a happy state of affairs but you mistakenly think that it is being brought about either somewhere else or by someone else or that it may be brought about by you in some other place, at some other time, or under some other circumstances, you are regrettably diminishing your own life, for this particular opportunity will never be yours again. But your loss is my loss as well, because if you keep your wits about you, you will not fail to affirm and to celebrate what is worth affirming and celebrating, thus helping me to come into my own. So do not fail to take charge of your own life, and do not miss the opportunity to judge the meaning and value of what you are doing or what is happening to you.

This does not mean that you will wallow in sheer joy and exuberance. To be fair to the total context of action, which is always contingent and never wholly under the control of the agent, the completing judgment must take into account objective circumstances. A realistic attention to circumstances will keep you from falling into narcissism or callousness, and will not tempt you into arrogance. Even altruism must be realistic if it is to be effective. Consider, for example, the following circumstances: What is the right thing to do for a person whose health is so impaired as to reduce his day-to-day existence to a constant struggle against pain? Or what can one do for someone suffering from a terminal illness? Although options are limited, something *can* be done. One may try to introduce as much peace, contentment, and joy into the sufferer's daily rounds as is possible. To the extent that these sorts of experiences enter their lives, their pain is at least made less unbearable and they are encouraged to focus attention on happenings that make their life worthwhile.

Openness to goodness need not diminish a person's sensitivity to evil. Indeed, a case can be made that only such sensitivity is likely to make a person more vigilant, unsentimental, and resolute in confronting evil. But there is no point in fighting evil unless the actualization of its opposite is possible. And one cannot be on the lookout for such actualization unless one has experienced some affirmations and celebrations, has judged some events to be worthy of benediction.

Self-evaluation is not always triumphant in every respect. For the most part human beings attain only a partial satisfaction of their needs, interests, and desires. No person is happy with all the decisions made in the past. Many choices leave traces of regret, remorse, or guilt. Some leave scars not only on one's psyche but also on the actual course of one's career. However, the awareness of one's own inadequacy, including instances of outright self-condemnation, is possible only because persons seek positive states of mind. Even to see oneself religiously as a sinful creature need not be a morbid psychological self-flagellation. The sense of sinfulness stems from the realization that the opposite is desirable. The state of sin is contrasted with the state of grace or blessedness. No one is to contemplate sinfulness neutrally, with a sense of equanimity. Sin is here not just to be punished; it is here to be overcome.

One might also say, more realistically, that the real sin is to *persist* in it. The sense of sinfulness may be seen as a self-directed perception of the desirability of reform. According to one plausible interpretation, the function of prayer is not to seek supernatural intervention in our external fortunes, but to redirect a believer's will and purposes. Likewise, the function of the sense of sin is to reorient oneself in the opposite direction. Unless there is a turning away from one's sinfulness, it is gratuitous to expect the intervention of grace; such an intervention would be undeserved. According to some tenets of Christianity, the grace bestowed on man is completely undeserved. In spite of that tenet, Christians are urged to *turn* their minds and hearts to Christ so that they become worthy of salvation in God's eyes. Man is worth saving because he has the capacity to condemn his own sinfulness. On the one hand, humility demands that one does not credit oneself with bringing about the act of grace. On the other hand, there must be a self-initiated response of the sinner to repent his ways for the act of grace to come off at all. The human soul must at least cooperate, or be interested in being in a state of grace. Thus, Christianity finds a germ of health in the capacity of persons to take an active part in the quest for salvation.

Preoccupation with sin, however, is not a sign of health. For a person to concentrate his gaze only on his scars and to allow them to dominate his self-consciousness is debilitating. A healthier course is to direct attention to activities and projects which allow one to move beyond the mistakes of the past. Past errors and transgressions need to be acknowledged and followed up by a real effort to repair the wrongs committed. But having done so, it is wiser to move on to activities and projects where you can function from your strengths and not from your weaknesses. You must judge which positive completions are still possible for you. It is one thing to learn from one's mistakes and to try not to repeat them, and quite another to be paralyzed and poisoned by them in whatever steps you undertake in the future. As long as you live, it is your task to make life a positive experience for yourself, a constructive force in the ongoing human enterprise, and a contribution to my cosmic pursuit of life values on earth.

I cannot predict how this pursuit will end. Whether it will go on indefinitely, discovering new possibilities, challenges, and wonders, or whether at some point it will meet its demise, conquered by the second law of thermodynamics, or, due to human error, or some catastrophic events, it is still up to persons to opt for rendering evaluative judgments on what they realize in their lifetimes. Every time you *can* exercise this option, it is a loss not to do so. Of course, the opportunity may not come every day, or it may be blocked for long stretches of time by tragic circumstances — illness, deprivation, poverty, oppression. But if no such external obstacles prevent you from living at the level of personhood and enjoying its prerogatives, you enrich your life, literally *give* it meaning and vibrancy, when you bless it, proclaim it to be good. As a person, alive and alert, you are in a position to give full cognizance to many varieties of goodness as they come to surface in you and in others. All persons are potential yes-sayers to life; indeed, yes-saying is their calling on earth. They cannot delegate this task to someone else or to some alleged power that will do it for them. And they should know that there is no greater satisfaction than the consciousness of having given to each experience its due, of having completed it by an affirming, understanding, appreciating act.

According to the Genesis story, when God made the world He found it very good. The act of benediction followed the act of creation. Since a deliberate introduction of values into the world is the special objective of persons, the account of Genesis is a perfect symbolic rendering of that objective, and, when universalized, it is applicable to every person's life. My cosmic spirit is aggrieved when the opportunities for benediction are squandered. To be a person is to be a cultivator of such opportunities; they are lost when the judgment on them is deferred to some other supposed authority. "The buck stops here!" Harry S. Truman reminded himself by placing this motto on his desk. This was Truman's way of making clear his special status as President of the United States, but this slogan is equally fitting for reminding all people that they preside over their personhood, over the decisions and judgments they are capable of making.

To be prepared to make confident value judgments is to live at the level of full awareness and alertness, optimally utilizing one's personal resources. *The human desire for the fullness of being, for having life and having it abundantly, expresses me in ways unparalleled in my entire cosmic span of space and time.* Persons are literally spokespersons for humanity, and ultimately for the whole of my reality, which includes all their concepts of God. For it is humanity itself that is my moving edge, enabling me to constantly replenish my thirst for new forms of excellence. By choosing and following their life plans, persons give expression and character to that edge. Since I have no predetermined goal toward which humanity is to aspire, each person helps to determine the nature of that goal. This is the insight behind Kant's claim that in choosing morally a person chooses for all rational beings and behind Dostoyevsky's slogan that everyone is responsible for everyone else.

To catch a glimpse of this insight is to sense at the same time a responsibility for one's life and the special privilege of being a person. Cosmic religiousness imbues a person with the conviction that each human life matters and matters absolutely. That conviction will be translated into serious efforts to develop the values of personhood in oneself, to confront resolutely circumstances which obstruct such development in others, and to contribute all one can to the prevention of a nuclear holocaust which threatens not only humanity but also other forms of life on earth. A person moved by cosmic religiousness can confidently declare on many occasions: "The values I am experiencing have absolute, irreplaceable worth because the world has never been *like this* before. In my person the cosmos comes to a bit of consciousness about itself and, through my ability to judge, gives me the right to say: life is very good."

Postscript

Looking back at the story I have put into the metaphorical mouth of the universe, I would like to summarize briefly the main concerns behind this admittedly unorthodox enterprise. Let me begin by repeating a disclaimer made in the preface: I do not pretend to have a privileged insight into the nature of things. The story presented here is accessible to anyone familiar with general facts about the universe and the restless creature inhabiting one of its tiny planets. The writer's motive was only to highlight some phenomena that in his opinion deserve special attention. In conclusion, let me focus on these phenomena in a more direct and summary way. I am hoping that the following observations about the world, its history and our place in it did not escape the reader.

The vivification of matter

Impressed with the immensity of the physical cosmos, we are devoting so much attention to the properties of "dead" matter in it that we fail to appreciate sufficiently the tremendous differences that were brought into the universe when a small proportion of matter on our tiny globe became vivified, life-bearing. The new quality or dimension I have characterized as meaning and value, suggesting that this dimension has made the universe much more interesting that it otherwise would have been. Indeed, in spite of its vastness, the physical universe is, in fairness, describable as literally dumb and dull. The organization, the inner structure and behavior of matter is highly intricate but relentlessly repetitive and rigid.

The personalization of life

Comparing the forms of life that gradually developed on earth, it is impossible not to notice the differences in their capacities and abilities. The dimension of meaning and value has reached a really full-blown stage with the advent of persons. The capacity to feel, characteristic of all life, when enhanced by intellectual processes, gives human beings a special status in the living kingdom. It enables them to transform their lives, individually and socially, into something highly sophisticated, refined, and unpredictable. The quality of experiences granted to the human form of life has lit up the otherwise silent and monotonous cosmic spaces with the spirit of knowledge, art, of open-ended creative enterprises. The scale of pain and pleasure, joy and sorrow, achievement and failure, good and evil, widsom and ignorance has widened in truly spectacular, unforeseen ways.

The substantivity of the moment

The temporal aspect of life has unjustifiably received bad press in most of our philosophies and religions. For complicated and partly understandable reasons, we tend to deplore the passage of time. And yet all meaningful and valuable events in the universe and in our human careers require time to *come* into focus. Timelessness is an enemy of meaning; nothing *happens* in eternity. Once more, I wish to quote Santayana's profound observation: "If time bred nothing, eternity would have nothing to embalm." This ironic comment directs our attention to the fact that the idea of meaning and that of birth, of generation, go together. The substance of the world is inherently temporal in the radical sense that only in time can any substance or any process acquire reality. Correspondingly, acts of appreciation and celebration, of realizing what is meaningful and valuable, are also temporal events. There is no goodness without goodness happening. Its substance is its occurrence.

The unconfinability of the spiritual quest

The title of this book suggests that the telling of the universe's story will reveal the possibility of a cosmic religion. Surveying human activities across the ages, it is impossible to overlook the element of aspiration that animates all creative effort. The aspirational nisus was immortally captured in Plato's philosophy, but it echoes and re-echoes in subsequent Western philosophies and religions. It is equally prominent in spiritual explorations of Eastern cultures. The proliferation of religious concepts and doctrines, and the views presented in several great world religions, testify to the urge to understand and to appreciate the universe holistically. But the constant ferment, debate, and controversy as to which religious world view is most defensible make it abundantly clear that the spiritual quest cannot be confined to what has already been thought or laid down. Religious piety can take the form of deep respect for this very open-endedness of the human spirit. That endeavor, however, will remain honest and healthy only when it does not abandon the critical capacities by means of which intellectually dubious and morally questionable beliefs and practices may be exposed. For that reason, it should not be up to theologians alone to determine who is and who is not religious or spiritually alive. "The spirit moveth where it listeth."

The accountability of persons

The new thing under the sun is the ability of human beings to act in the light of acquired standards and to change their environment in terms of what has been discovered to be valuable, worth bringing about. The special characteristics of personhood have come into being in the context of community and cooperation, and the continuing flourishing of persons depends on mutual cherishing of the complementarity of human excellences. As originators of ideas and organizers of common efforts, persons cannot escape responsibility for the results of their actions. Indeed, personhood and accountability mutually imply each other. Accountability has greatly expanded with the advent of scientific knowledge, which affects the range of possible human influence on the fate of the earth and of its inhabitants. Responsibility cannot be delegated to impersonal forces. If humanity becomes impersonal or intolerant, self-indulgent or lazy, callous or indifferent, skeptical or defeatist, it runs the danger of returning the universe to a state devoid of civilization, culture, and adventures of ideas. These qualities, as far as we know, are rare guests in interstellar spaces. It is our job to make them feel at home.

Notes

[1]Annie Dillard, *Pilgrim at Tinker Creek,* New York: Harper's Magazine Press, 1974, p. 64.

[2]Stephen Lackner, *Peaceable Nature,* San Francisco: Harper & Row, 1984, p. 26.

[3]Alexander Solzhenitsyn, *The Gulag Archipelago,* Thomas P. Whitney, trans., New York: Harper & Row, 1974, Vol. I., p. 168.

[4]Joseph P. Lasch, *Helen and Teacher: The Story of Helen Keller and Anne Sullivan Macy,* New York: Delacorte Press, 1980.

[5]Kenneth L. Patton, *The Religion of Realities,* Ridgewood, NJ: The Meeting House Press, 1977, p.4.

[6]Ibid., p. 10.

[7]G.W. Hegel, *Lectures on the Philosophy of Religion,* Speirs and Sanderson, trans., London: Kegan Paul, 1898, Vol. I, p. 228.

[8]Friedrich Nietzsche, *Thus Spoke Zarathustra,* Walter Kaufmann, trans., New York: Viking, 1954, p. 121.

[9]David L. Norton, *Personal Destinies,* Princeton University Press, 1976.

[10]Stephen Lackner, Op. cit., p. 136.

[11]Aristotle, *Politics,* 1260a.

[12]Alexander Solzhenitsyn, *The First Circle,* Thomas P. Whitney, trans., New York: Bantam Books, 1968, pp. 451-52.

[13]Jonathan Schell, *The Fate of the Earth*, New York: A. Knopf, 1982, p. 178.

[14]Ibid., p. 174.

[15]Ibid., p. 177.

[16]Lewis Thomas, *Late Night Thoughts on Listening to Mahler's Ninth Symphony,* New York: Viking, 1983, p. 166.

[17]Joseph Campbell, "The Secularization of the Sacred," *The Religious Situation,* Donald R. Cutler, ed., Boston: Beacon Press, 1968.

[18]J.W. Goethe, *Faust,* B.Q. Morgan, trans., New York: Liberal Arts Press, 1954, p. 42.

Index

accountability, 119, 131
activism, 118
affirmation, 9, 42, 86, 126
aggressiveness, 31
altruism, 33
Anaxagoras, 19, 20, 116
anthropomorphism, 21
Aristophanes, 73, 79
Aristotle, 50, 75, 109, 116
Armstrong, Neil, 53
autonomy, 29, 38-47, 58, 67, 94

Bacon, Francis, 15
Beard, Charles A., 101
beauty, 11
Beethoven, Ludwig van, 53, 60
benediction, 62, 126
Bentham, Jeremy, 59
Berkeley, George, 42
blessedness, 58, 62-64, 67, 117
Bruno, Giordano, 51

Campbell, Joseph, 111-112
capitalism, 87, 103
caring (care), 11, 30-32, 34, 53
character, 39

Christianity, 29, 94, 106, 109, 110-113,
    119, 125
civilization, 17, 51-53, 104, 131
Clifford, W.K., 54
Collingwood, R.G., 115
Columbus, Christopher, 53
communication, 91, 93
communism, 90, 93, 103
community, 79, 84-86, 91, 94, 131
community of interpretation, 72-73
conatus, 4-8, 10, 31, 33, 37, 47, 74
concepts, 47-48, 52
cooperation, 12, 31, 38, 69, 93, 131
Cousins, Norman, 61
creation, 108
creativity, 50, 130
crime, 36-38
cruelty, 37
Curie-Sklodowski, Marie, 53

Darwin, Charles, 7
democracy, 87, 89
Democritus, 54
Descartes, René, 29
Dewey, John, 121
Dillard, Annie, 6

distraction, 43-44
Dostoyevsky, Fyodor, 53, 127

Eckhart, Meister, 57
ecology, 31
education, 60-61, 104
Einstein, Albert, 53
equality, 79, 86-87
eternity, 130
ethnicity, 80-84, 96
ethnocentrism, 81-84
eudaimonia, 59, 62
evil, 37-38, 125, 130
evolution, 10, 21
existentialism, 42, 64-66
experience, 41, 52, 117
explanation, 20
extinction, 7, 97-104

fanaticism, 37, 40
feeling, 2-4, 25, 27, 31, 47, 130
form of life, 17, 19, 27, 30-31, 52, 130
freedom, 30, 42, 87, 89-90
Freud, Sigmund, 44-45

genes, 7
genius, 40-41
God, 54, 57, 82, 107-119, 121, 123
Goethe, Johann W., 12, 53, 78, 117-
    119
goodness, 11, 19, 25, 67, 125, 127

Hammurabi, 51
happiness, 58-62, 67
Hebbel, Friedrich, 77
Heidegger, Martin, 11
Henry, Patrick, 51
Hinduism, 120
Hobbes, Thomas, 37
honesty, 38
humanity, 12-22, 52-53, 58, 66, 127,
    130
Hume, David, 42

ideals, 123
ideology, 86-90, 92, 94

imagination, 50, 106
immortality, 108, 116-119
individuality, 23-30, 67, 85, 95-96, 124
instinct, 12-13, 15-16, 21-23, 31
integrity, 38-47, 58, 67
intelligence, 12, 17, 48, 51
intelligence, artificial, 19-20, 23
intentionality, 53

Jefferson, Thomas, 51
Joan of Arc, 51
judgment, 25, 27, 30, 68, 79, 113-116
justice, 54, 87

Kant, Immanuel, 29, 34-35, 52-54, 116-
    118, 127
Keller, Helen, 44
King, Martin Luther, Jr., 82, 96
knowledge, 13, 24, 26, 47, 50, 62
Kropotkin, Peter, 12
Kuhn, Thomas, 65

Lackner, Stephen, 10-11, 63
language, 13-18, 24, 31, 47, 55
Lincoln, Abraham, 51, 53
life, 3-11, 21, 30-31, 129
Locke, John, 42, 52
love, 76-77, 79, 110-112

Marxism, 86-87, 90
matter, physical, 1-2, 21, 31, 48, 129
meaning, 4, 47-58, 66, 97
Mill, J.S., 29, 59
Milosz, Czeslaw, 23
miracle, 4, 123
monotheism, 82, 107, 109
morality, 30-38, 77-79, 116-117
Mother Theresa, 96
mysticism, 47, 56-58, 63, 67

nature, 5-6, 27, 30-32
Nelson, Admiral, 51
Newton, Isaac, 65
Nietzsche, Friedrich, 24, 29, 38, 57
Norton, David L., 58-59
nuclear danger, 98-104, 127

objectivity, 31, 33, 53-54
obligation, 55
organism, living, 3-4

pain, 8-9, 21
Pascal, Blaise, 64, 103
patriotism, 69
Patton, Kenneth, 49-50
perception, 16, 53
person, 22-30, 50, 127
personhood, 25, 29, 36, 45, 57-58, 67,
    80, 86, 99, 103, 122
piety, 53-56, 67
Plato, 51-52, 73, 75, 107-108, 130
pleasure, 8-9, 21, 27
progress, 69
protoplasm, 3
prudence, 33-34
purpose, 7, 8, 29

quasi-moralities, 35

rationality, 15-17, 19-20, 33, 99
redemption, 112-113
reference, 13-14, 24, 42, 47
relativism, 65
religion, 105-108, 130-131
 religion, common core of, 119-122
religiousness, 121-127
reproduction, 5
responsibility, 31-33, 89, 100-101, 114,
    127, 131
rights, 55
Rilke, R.M., 52, 76-77
Royce, Josiah, 72
rules, 27

Sagan, Carl, 53
Saint Paul, 75-76
Sakharov, Andrey, 91
Santayana, George, 130
Sartre, Jean-Paul, 4, 42-43, 80
Schell, Jonathan, 99-100
Schopenhauer, Arthur, 26, 39
Schweitzer, Albert, 55-56
scepticism, 21, 44-46

science, 48-50, 65-66, 94, 104, 131
self, 42-43
self-appraisal, 41
self-conception, 41, 58-59
self-construction, 46-47
self-deception, 41, 45
self-development, 39-40
self-interest, 33
self-reference, 24, 27, 42, 47
sentience, 5, 26, 31, 44, 47, 55
sex differential, 73-80
sexual roles, 74, 81
Shakespeare, 53, 63, 115
sin, 125-126
Spinoza, Baruch Benedict, 4, 56
social determinism, 70-73
social identity, 24-25
socialism, 87, 90-91
socialization, 71-72
society, 70-71
Socrates, 19-20, 51, 54, 56, 111
Solzhenitsyn, Alexander, 37-38, 85-88,
    95-96
Sophocles, 53
standards, 15, 20, 24-27, 32, 38, 49,
    51, 99
Stoicism, 94, 107
supernatural, 123, 125
survival, 10, 31
symbiosis, 12, 31, 34

temporality, 61-62, 130
Thomas, Lewis, 102
Tillich, Paul, 106, 121
Tolstoy, Leo, 59
Toscanini, Arturo, 60
totalitarianism, 87
tribalism, 68-70, 88
Truman, Harry S., 126-127
truth, 45, 54
truthfulness, 45, 54

ultimate concern, 105-106
unconscious motives, 44-45

value, 8-9, 16, 20-22, 25, 27, 30-31, 40,
    51, 58, 116, 123

vegetarianism, 55
virtue, 117

Whitehead, A.N., 116, 121
Wittgenstein, Ludwig, 16-17
wisdom, 63, 130

IMAGE SPIRITUAL CLASSICS

Available at your local bookstore, or if you prefer to order direct, use this coupon.

ISBN	TITLE	PRICE	QTY	TOTAL
19614-8	Sadhana by Anthony de Mello *Eastern Meditations in Christian Form*	$7.95 ×	_____ =	_____
02861-X	The Imitation of Christ by Thomas à Kempis *The Classic Spiritual Work*	$6.95 ×	_____ =	_____
14803-8	The Wounded Healer by Henri J. M. Nouwen *"Nouwen at his best"*	$6.95 ×	_____ =	_____
12646-8	Apologia Pro Vita Sua by John Henry Cardinal Newman *One of the Most Influential Books of Western Civilization*	$9.95 ×	_____ =	_____
17034-3	Breakthrough by Matthew Fox *A Brilliant Study of Creation Spirituality*	$15.00 ×	_____ =	_____
02955-1	The Confessions of Saint Augustine Translated by John K. Ryan *The Greatest Spiritual Biography of All Time*	$6.95 ×	_____ =	_____

Shipping and handling (Add $2.50 per order) _____

TOTAL _____

Please send me the titles I have indicated above. I am enclosing $ _____ . Send check or money order only (no CODs or cash please). Make check payable to Doubleday Consumer Service. Allow 4-6 weeks for delivery. Prices and availability subject to change without notice.

Name: _____

Address: _____ Apt # _____

City _____ State _____ Zip _____

Send completed coupon and payment to:
Doubleday Consumer Service
Dept. Z-109
P.O. Box 5071
Des Plaines, IL 60017-5071

9-91/Z-109